Neuromorphic Engineering

Neuromorphic Engineering

The Scientist's, Algorithm Designer's, and Computer Architect's Perspectives on Brain-Inspired Computing

Elishai Ezra Tsur
The Open University of Israel

CRC Press is an imprint of the
Taylor & Francis Group, an **informa** business

First edition published 2022
by CRC Press
2 Park Square, Milton Park, Abingdon, Oxon, OX14 4RN

and by CRC Press
6000 Broken Sound Parkway NW, Suite 300, Boca Raton, FL 33487-2742

© 2022 by CRC Press.

This book is based on academic research held at the Open University of Israel and its publication is under license by agreement with the Open University of Israel.

CRC Press is an imprint of Informa UK Limited.

Reasonable efforts have been made to publish reliable data and information, but the author and publisher cannot assume responsibility for the validity of all materials or the consequences of their use. The authors and publishers have attempted to trace the copyright holders of all material reproduced in this publication and apologize to copyright holders if permission to publish in this form has not been obtained. If any copyright material has not been acknowledged please write and let us know so we may rectify in any future reprint.

Except as permitted under U.S. Copyright Law, no part of this book may be reprinted, reproduced, transmitted, or utilized in any form by any electronic, mechanical, or other means, now known or hereafter invented, including photocopying, microfilming, and recording, or in any information storage or retrieval system, without written permission from the publishers.

For permission to photocopy or use material electronically from this work, access www.copyright.com or contact the Copyright Clearance Center, Inc. (CCC), 222 Rosewood Drive, Danvers, MA 01923, 978-750-8400. For works that are not available on CCC please contact mpkbookspermissions@tandf.co.uk

Trademark notice: Product or corporate names may be trademarks or registered trademarks and are used only for identification and explanation without intent to infringe.

Library of Congress Cataloging-in-Publication Data

ISBN: 978-0-367-67680-3 (hbk)
ISBN: 978-0-367-69838-6 (pbk)
ISBN: 978-1-003-14349-9 (ebk)

Typeset in LMR12 font
by KnowledgeWorks Global Ltd.

To my son Ben and my daughter Maya,
for teaching me emergent love

"There will one day spring from the brain of science a machine or force so fearful in its potentialities, so absolutely terrifying, that even man, the fighter, who will dare torture and death in order to inflict torture and death, will be appalled, and so abandon war forever." Thomas A. Edison.

Contents

About the author

Elishai Ezra Tsur
Neuro-Biomorphic Engineering Lab
Open University of Israel

Elishai is the principal investigator of the Neuro-Biomorphic Engineering Lab (NBEL-lab.com) and an Assistant Professor for Mathematics and Computer Science at the Open University of Israel. In his research, Elishai studies the realm of brain-inspired machines. He utilizes artificial brains to develop new frameworks for robotics and vision processing. Elishai is a university lecturer for artificial intelligence, mathematical modeling, software engineering, and computational biology. He was named an INK Fellow in 2013 (INK, India) and a Dean Fellow for Computational Neuroscience in 2018 (Weizmann Institute of Science, Israel). Elishai holds degrees in Life Sciences (BSc), Philosophy and History (BA), Computer Science (MSc), Bioengineering (MSc, PhD), and Computational Neuroscience (Post Doc).

Preface

The brain is not a glorified digital computer. It does not store information in registers and it does not mathematically transform mental representations to establish perception or behavior. The brain cannot be downloaded to a computer to provide immortality nor can it destroy the world by having its emerged consciousness traveling in cyberspace. In his article: *"The Empty Brain"* in the *Aeon Magazine* in 2016, *Robert Epstein* outlines his arguments against the Information Processing (IP) paradigm (or the mind-computer metaphor) arising from the *"prime facie digital"* of the brain. Analogously to the hydraulic model of cognition, which persistently argued for 1,600 years for having "humors" accounted for mental capacity, the IP paradigm plainly reflects our current state-of-the-art machinery. However, studying the brain's core computation architecture can inspire scientists, computer architects, and algorithm designers to think fundamentally differently about their craft.

Neuromorphic engineers have the ultimate goal of realizing machines with some aspects of cognitive intelligence. They aspire to design computing architectures that could surpass existing digital von Neumann-based computing architectures' performance. In that sense, brain research bears the promise of a new computing paradigm. As part of a complete cognitive hardware and software ecosystem, neuromorphic engineering opens new frontiers for neuro-robotics, artificial intelligence, and supercomputing applications.

This book will present neuromorphic engineering from three perspectives: the scientist, the computer architect, and the algorithm designer. We will zoom in and out of the different disciplines, allowing readers with diverse backgrounds to understand and appreciate the field. Overall, the book will cover the basics of neuronal modeling, neuromorphic circuits, neural architectures, event-based communication, and the neural engineering framework. Readers will have the opportunity to understand the different views over the inherently multidisciplinary field of neuromorphic engineering. The book aims to give the reader a wide description of neuromorphic engineering. It will, therefore, not dive into many of the technical intricacies of the discussed materials. Mathematical proofs will be detailed only when the discussed topics are fundamental to understanding the field's core aspects. Therefore, when exact mathematical formalism is not appropriate for the typical reader's background or expertise (e.g., derivation of the nFET sub-threshold transfer function), the intuition of proofs will be given and the interested reader will be referred to other specialized textbooks. The book will walk on the thin line of being descriptive enough to give the reader a clear understanding while not being too technical and tedious.

This book revolves around different aspects of engineering and biology. However, the field of neuromorphic engineering has deeper roots, profoundly embedded in man's innermost passion and fear from the creation of artificial conscious life. In Jewish history, it is known as "the golem". To enrich the book with another perspective, which traditionally has nothing to do with engineering and computing, I asked Prof. Tzahi Weiss to write a foreword on the Jewish tale's history about the artificial man. While delving into the intricacy of the mathematical models described later in this book, we should be able, from time to time, to turn a few pages back and recollect the passion and the fear which have enriched our cultural life and are fundamentally anchored to the field of neuromorphic engineering.

This book follows five years of teaching the course: *Brain-Inspired Computing* at the Weizmann Institute of Science, The Interdisciplinary Center, and the Hadassah Academic Center. The book was designed for graduate students in Computer Science Electrical Engineering, Neuroscience, and Computational Biology. Some familiarity with the fundamentals of Object-Oriented Programming (OOP), machine learning, algorithms, computer architecture, parallel computing, electrical circuits, and neuroscience is desirable but not required.

I would like to thank the students of the Neuro-Biomorphic Engineering Lab (NBEL-lab.com) at the Open University of Israel. The discussions I had with my graduate students throughout the nights and into the mornings fueled the depth and breadth of my appreciation for the field. The opportunity to teach neuromorphic engineering at the departments of computational neuroscience, neurobiology, and computer science ignited my passion for writing this book. I would also like to extend my appreciation to my colleagues at the Open University of Israel and particularly to Prof. Tzahi Weiss. I would like to thank Prof. Chris Eliasmith from the University of Waterloo, Canada, who inspired me to walk through the intricacies of neuromorphic engineering and permitted the utilization of his teaching methodologies for the description of the neural engineering framework in this book. Finally, I would like to thank the Open University Publishing House and the CRC production team for bringing these words to light.

Elishai Ezra Tsur

Foreword: a tale about passion and fear

By Prof. Tzahi Weiss. Prof. Weiss is a professor of Jewish Thought and Hebrew Literature in the Department of Literature, Languages and the Arts at the Open University of Israel. He is the author of four books, most recently, "Sefer Yesirah and its Contexts: Other Jewish Voices" (Penn Press 2018).

A SHORT HISTORY OF THE JEWISH TALE ABOUT THE ARTIFICIAL MAN, "THE GOLEM"

Introduction

The second account of the creation of the world by God in the book of Genesis is a very interesting narrative. Told from a psychological point of view with regard to the relations between God and human beings, it suggests that God harbored a certain fear of his own creation. God who forms Adam from dust and afterwards Eve from one of Adam's ribs subsequently warns them that they are not to eat from the tree of knowledge giving no reason or clarification for this ban. Nevertheless, the text does offer an explanation expressed by the serpent: "For God knows that when you eat from it your eyes will be opened, and you will be like God, knowing good and evil" (Gen. 3:5). This explanation discloses an imminent threat and bespeaks a deep fear: eating from the tree of knowledge could engender God-like humans and God might thereby forfeit his inimitable transcendence over humans. The story of God, Adam, Eve, the serpent and the tree of knowledge differs fundamentally from the model of good parenting. In raising and educating our children we make the tree of knowledge available to them, they eat its fruits incorporating the knowledge at our disposal and becoming with time independent and autonomous. This development, of course, can also arouse psychological tensions between parents and children: anger, envy and even death wishes. Nevertheless, the model described in the book of Genesis differs basically from that of parenting: God created a new species – Humans are not God. The threat they pose to (overthrow) their creator lies in the fact that they are not meant to be God but are rather “like God”. In the context of this book one might say that God in the second tale of creation is more like an inventor of a non-human intelligence who cannot predict the consequences of his innovation, than a parent to a child. The legend of the golem, one of the most famous Jewish narratives, is perhaps more similar in that sense to the position of God in the book of Genesis, the tension between the passion to create and the dread that is thereupon ignited by the lack of control over the consequences. The first formulation of the tale of the golem is to be found in the Talmud which was edited around the middle of the first millenium CE, and later evolved in Jewish sources during the Middle-Ages and the Modern-Era acquiring many and diverse versions. In the 19th century, it ascended to the height of its circulation reaching its peak in the 20th century at which time it also spread to many non-Jewish sources. The majority of modern sources from the mid 19th century on describe the creation of this artificial anthropoid by R. Judah Loew ben Bezalel Halevi from Prague (d. 1610) known as the Maharal of Prague.

In what follows I will depict four different stages in the evolution of the tale about the golem each of which will exemplify how the tension between the passion to create an artificial man and the concomitant fear of what the outcome and repercussions this created anthropoid might ensue in fact molded the story narrative. In the first part of the paper, I will discuss the tale about the golem in sources from late antiquity and the middle-ages in which the Jewish sages, though eager to create an artificial human are yet afraid that this will be an act of rebellion against God as “the creator”.

In the second part, I will analyze texts from the later middle ages that present instructions for the creation of a golem, but express fear and caution that inaccurate use of those prescriptions may cause terrible damage. In the third part I will brings stories from the early modern period about a creation of golem as a servant and the fear that he will act against his masters. In the four part I will discuss a less known account about the creation of a female golem. As we will see, in this instance the male creators wanted to create a female for their own needs but were afraid to realize their desires toward the female they had created.

To those engineers who are to be the readers of this book, I hope that my contribution as an historian of ideas in presenting these versions of the golem's legend synchronically will be to demonstrate that the apprehension concerning the creation of different sorts of artificial intelligence has a long history.

"What would be the outcome"? - The early Jewish tales about the creation of an artificial man

The Talmud is the series of Jewish tractates that documents rabbinic discussions about the Jewish commandments as well as legends about biblical figures and tales about Jewish sages. Those tractates were written and edited by Jewish rabbis from Palestine and Babylon between the 3rd and the 7th centuries CE. In one of the Talmudic tractates named Sanhedrin, a sage named Rava makes an astonishing declaration according to which, there is no difference between the righteous and God, and if the righteous only did wish they could even create the world:

> Rava said: If the righteous desired it, they could create the world, for it is written: (Isa 59:2) "But your iniquities have separated between you and your God". (b. Sanhedrin 65b)

Rava says that since the main matter which separates men from God is human transgressions, the pious are near to God and as such, if they desire, they can create the world. After the declaration of Rava, the redactor of the Talmud inserts a tale about the creation of an artificial man by the very same Rava:

> Rava created a man. He sent him to R. Zera. [R. Zera] spoke to him and [the artificial man] did not answer him. He [R. Zera] told him: are you from the group (hevraya)? Return to thy dust! (b. Sanhedrin 65b).

According to the tale, Rava created an artificial man, and sent it to his friend, R. Zera. R. Zera knew this to be an artificial man and tried to talk to him, but did not receive any answer. R. Zera understood that the golem is an artificial man. According to most of the interpreters of the tale 'the group' (hevraya), a nickname for magicians and in this case for a magical act performed by Rava. Therefore R. Zera tells the golem to return to his dust. Like many rabbinic tales, this tale about the golem is articulated in a condensed manner containing very few words. Nonetheless, one can notice the dissonance between Rava's conceited declaration and the fact that the man that he created could not even speak. It was not too hard for R. Zera to understand that this is not a real man. From a literary point of view due to the use of juxtaposition, the first known Jewish tale about a creation of an artificial man acquires an ironic tone that says: even a notable rabbi like Rava cannot in fact create a man and there is an unbridgeable gap between God and even the most pious. Rava's declaration in reality turns out to be no more than a fiasco.

The Talmudic tale about the golem is extremely short, but around the 11th century CE, one finds a more detailed depiction of this tale. This elaborate version of the tale attributes the creation of the golem to other figures: Jeremiah the prophet and his legendary descendent Ben-Sirah. According to this new medieval narrative, both created an artificial man by learning a Jewish composition named Sefer Yesirah, 'the Book of Formation'. This time they succeeded in the creation of the artificial man. Nevertheless, the dummy rose against its maker:

> And Ben Sira, [...] went to Jeremiah, and they studied it for three years and understood it, and a man was created before them, and on his forehead was written "And the Lord, God, is Truth" and in his hand was a knife and he erased the letter alef from emeth. Jeremiah said: "Why are you doing that? Can the Truth not be?" He said to them: "I will tell you a parable. What does this resemble? [It resembles] a man who was a builder and was wise. When the people saw him, they made him their king. A few days later, other people came and learned that craft, and [the people] abandoned the first one and followed the others. Likewise, the Holy One, blessed be he, may his name be exulted [...] created the world, and they made him king over his created beings. Yet now that you have come and have done the same thing, what would be the outcome? Everyone would leave him and would follow you. What would become of him who created you?" They said to him: "In that case, what should we do?" He answered: "Turn him [i.e., me] back." And that man became dust and ashes. (Translation according to Weiss, Sefer Yesirah, p. 94)

This version of the tale has an interesting twist as the golem after being created by Ben Sirah and Jeremiah admonishes them saying that such an act of creation is in fact a rebellion against God and therefore he should be returned to dust. According to the tale, on the golem's forehead the Hebrew words from the book of Jeremiah: "And the Lord, God, is true" [Jer. 10:10] are written. The golem comprehends that the very fact that he was created entailed a theological rebellion, decides to erase the letter aleph from the word Emeth, which turns the word from truth to the word "dead". Now on the golem's forehead it is written "And the Lord, God, is dead"! Ironically, it is the golem who turns out to be the wise person in this legend and he himself elucidates the reasoning which in the end brings about his own destruction, as a metaphysical and moral issue - an act which upholds and protects the theological order. God is not only the creator, he is the order of the world. The golem asks they who made him if they understand the full consequences of their deeds, in his words: "what would be the outcome?" I think that these words put in the mouth of the golem by the narrator express and formulate the deep fear attendant upon an innovator who is eager to create, but cannot predict the outcomes of the realization of his ideas.

The fear of fatal error, the medieval variant of the golem

From the 12th century CE onward Jews in western Europe developed new genres of mystical, magical and mythical writings. In the south of what is today considered as France, in Provence, and in the north part of the Iberian Peninsula, in Catalonia and north Castile, the well-known genre named Kabbalah was being developed. At approximately the same time in Germany of today and north-east France a different mystical genre of writing developed by a group of people named 'The German Pietists'(hasidei ashkenaz). In the writings of hasidei ashkenaz one can find for the first time actual prescriptions for the creation of an artificial man. As we have seen in the Talmud as well as in the later tale about the golem, the depiction of its creation is merely legendary; these literary narratives are mainly about the Jewish sages or biblical figures who created a golem. In the case of the 12th–13th centuries texts from north-western Europe we find more technical instructions concerning the process of the creation. Thus, for example, at the end of his commentary to Sefer Yetsirah Elazar of Worms, one of the most prominent German pietist, writes a short prescription for the creation of a golem:

> Whoever studies Sefer Yetsirah has to purify himself [and] don white clothes [...] It is incumbent upon him to take virgin soil from a place in the mountains where no one has plowed. And he shall knead the dust with living water, and he shall make a body [golem] and shall begin to permutate

> the alphabets [...] each limb separately, each limb with the corresponding letter mentioned in Sefer Yetsirah. And the alphabets will be permutated at the beginning, and afterwards he shall permutate with the vowel [. . .] And always, the letter of the [divine] name with them, and all the alphabet [...] He shall do all this when he is pure [...]. (translations according to Idel Golem, p. 56).

We have already mentioned Sefer Yetsirah, the Book of Formation. This short treatise depicts the creation of the world by means of the decimal counting system and the twenty-two letters of the Hebrew alphabet. According to Sefer Yetsirah each letter corresponds to a limb in the human body. Since the 11th century at the latest, Jewish sages considered Sefer Yetsirah to be the manual for the creation of an artificial man. In this text, Elazar of Worms depicts precisely how to create a golem by using the correspondence between letter and limbs in Sefer Yetsirah, correctly. According to Elazar of Worms, one should take soil "from a place in the mountains where no one has plowed And he shall knead the dust with living water and he shall make a body [golem]" After the creation of the body it is necessary to enunciate the letters of the alphabet and to create the golem in purity. Historically, this is a new way of relating to the creation of a golem - as an actual act and not only as a literary narrative. The prescription by Elazar of Worms is not unique in the writings of 'The German Pietists'. Among these concrete instructions for the creation of a golem, one can find some which contain alongside the prescription also expressions of severe caution whose main motivation is to deter people from making a golem. An example for that is to be found in a commentary to Sefer Yetsirah that is attributed to R. Saadia Gaon a 10th century rabbi who lived in Babylonia, but was actually written by sages of one of the branches of the German pietists, named: 'The Unique Cherub Circle', who were active in what is currently the north of France . This text serves as a good example of the manner in which the passion to create a golem can be accompanied with a deep fear of a fatal error, which might occur in the process:

> There are persons who explain [the words] "the wheel going frontward and backward," [=form Sefer Yetsirah] that the Creator has given power to the letters, [so that when] someone creates his creature out of virgin soil, and he kneads it and buries it in the soil, and makes a circle and wheel around

> the creature and says at each and every circumference one alphabet [...] And if he goes forward, the creature rises to life by the power of the utterance of the letters, since God gave them power. If he wants to destroy what he created, he returns backwards [going] around pronouncing the letters, and the creature will sink by itself and it shall die. So it happened, once to R. Y. ben A [=unrecognized rabbi] and his disciples who were studying Sefer Yetsirah. They wanted to create a creature but they erred in [the direction] of their walking, and they went backwards until they sank in the earth up to their navels, by means of the letters. They were not able to go out, and they screamed. R. Y. ben A. heard their voice and told them, "Say the letters of the alphabets going forwards, just as you went backwards." They did so and went out [. . .] (Translation according to Idel Golem, pp. 81–82)

It seems, therefore, that as the tale about the golem received a more concrete form, a recipe to create him, the fear of an error in the process itself became in turn more concrete. The urge to create an artificial man and the consequent fear which is aroused do not disappear, they just change their shape.

The dummy rose against its maker: the golem from the early modern period onward

The creators of a golem, as we have depicted them, in the last two sections did not assign any definite purpose to the golem. In the Talmud the making of the golem was a demonstration of the high abilities and qualities of the sages comparing them to God, a comparison that was shown to end as a total fiasco. In the texts written in the circles of the German pietists, the creation of a golem was in and of itself the main goal as well as the main danger, yet again the golem as a literary figure has neither content nor purpose. From the early modern period onward, unlike the above mentioned texts, making the golem has a concrete goal: the golem is designed to be the servant of his makers and their community. The literary figure of the golem developed more and more throughout the years and acquired an interesting and rounded character from the 20th century onwards. One of the earliest descriptions of the golem as a servant, was documented by Christoph Arnold a 17th century non-Jewish folklorist. According to the tale of Arnold, which has equivalents in contemporary Jewish texts, a famous Jewish rabbi and magician named Elijah of Chelm (d. 1583) created a golem that was intended to be his servant. The golem got bigger and bigger each day, and became a danger to his maker who could not reach the golem's forehead to erase the letter aleph and return him to his dust. The rabbi had a tricky idea: he asked the golem to remove his boots and then he erased the letter aleph from his forehead. The plan went well. Nevertheless, when Elijah erased the letter the golem fell on Elijah and crushed him:

> After saying certain prayers and holding certain fast days, they make the figure of a man from clay, and when they have said the ineffable name (shem hamphorash) over it, the image comes to life. And although the image itself cannot speak, it understands all that it is told and commanded to do; among the Polish Jews it does all kinds of housework, but is not allowed to leave the house. On the forehead of the image, they write: emeth, that is, truth. But, an image of this kind grows each day; though very small at first, it ends by becoming larger than all those in the house. In order to take away his strength, which ultimately becomes a threat to all those in the house, they quickly erase the first letter aleph from the word emeth on his forehead, so that there remains only the word meth, that is, dead. When this is done, the golem collapses and dissolves into the clay or mud that he was. [...] They say that a baal-shem [=magician] in Poland, by the name of Rabbi Elias, made a golem who became so large that the rabbi could no longer reach his forehead to erase the letter aleph. He thought up a trick, namely that the golem, being his servant, should remove his boots, supposing that when the golem bent over, he would erase the letters. And so it happened, but when the golem became mud again, his whole weight fell on the rabbi, who was sitting on the bench, and crushed him. (The translation is according to Scholem, the Golem, pp. 200–201)

This is an early text about the danger of using the golem as a servant, but it exemplifies very well the literary structure of making a golem; one can only create a golem but cannot control the ramifications of one's own creation. There is no doubt that Rabbi Elijah is wiser than the golem, and his trick demonstrates that. Nonetheless, the golem can kill his creator. The message of this tale to the reader is to be aware of the danger in magic and specifically in making an artificial man. Such a servant might kill his master and making a golem can turn out to be more of a danger than beneficial.

The female golem and the fear of sexuality

A lesser known version of the golem's legend describes the creation of a female golem. An example for that can be found in the writing of the 16th century kabbalist, Moses Cordovero (d. 1570). Cordovero interprets the defamation of Joseph to Jacob against his brothers in the book of Genesis, as a misunderstanding that is precipitated by the creation of a female golem. Cordovero presumes that both Joseph and his brothers are righteous. If so, he asks how is it possible that Joseph informs on his brothers to his father, someone in this story, either Joseph or the brothers did wrong. As a solution to this problem, Cordovero suggests that the brothers created a female golem and had some kind of sexual interaction with her. Joseph did not know that 'she' was an artificial woman and reported to his father that his brothers did something inappropriate. Nonetheless, his bothers in fact did no wrong since a relationship with an anthropoid has no significance according to the Jewish law. The status of the female golem in this story is not clear but it seems that Cordovero indicates that at the root of this misapprehension there lies an important issue dealing with appearance and attribution. While the female golem is not a real woman, it is a fact that Joseph perceived her as a real person and in a way so did his brothers who had a sexual motivation in creating her. A more complex expression of the tension between the desire and fear of a female golem can be found in one of the short stories of the Israeli Noble prizewinner S.Y. Agnon (1887–1970). In one of his most studied yet enigmatic stories named 'Ido and Enam' a mysterious relationship between a scholar named Gideon Ginat and a woman named Gemolah develops. The Hebrew letters of the name Gemolah and 'the golem' are identical and indeed the narrator of the story says that Ginat, like a medieval Jewish rabbi, Samuel ibn Gabirol, created an artificial woman. Nevertheless, at the end of this story when Gemolah comes to Ginat's apartment and asks him to be with her, he refuses and sends her from him and it seems that he is afraid of the straightforward manner in which she expresses her wish to be with him:

> Gemulah said, 'After all the years that I have waited for you! Now you say, "Go, Gemulah." [...] And you, Hacham

> Gideon,' said Gemulah, 'what are you to me?', 'I am nothing.' Gemulah laughed. 'So you are nothing! You are a good man, you are a lovely man, in all the world there is no man so good and so lovely as you. Let me stay with you [...] The young man said, 'Go, Gemulah, go [...]' (Edo and Enam, pp. 221–222)

A few concluding words

The Jewish legend about the golem is a powerful cultural narrative that has persisted for more than 1,500 years. During that long period it has been articulated in many ways each of which reflects its own cultural context. The main goal of this short essay was to demonstrate that beyond the differences between the versions of the golem's legends there is a stable core that reflects the desire or even the passion to create something highly innovative and the various attendant fears which may be connected to this very same achievement. I believe that in this respect, the passion and dedication of modern scientists to find the innovative ideas which might improve the world occupies the same psychological structure that we have seen in the different versions of the golem tale: the urge to create is accompanied and tempered by the dread of the dangers and consequences of their cutting edge ideas.

Selected bibliography on the golem and texts that were mentioned in the text:

- S.Y. Agnon, "Edo and Enam" in: Two Tale by SY Agnon, Walter Lever (trans.) New York: Schocken Books, 1966, 143–232
- Maya Barzilai, Golem: Modern Wars and Their Monsters, New York: NYU Press, 2016
- Cathy S. Gelbin, The Golem Returns: From German Romantic Literature to Global Jewish Culture, 1808–2008, Ann Arbor: The University of Michigan Press, 2011
- Arnold Goldsmith, The Golem Remembered, 1909–1980: Variations of a Jewish Legend, Detroit: Wayne State University Press, 1981
- Moshe Idel, Golem: Jewish Magical and Mystical Traditions on the Artificial Anthropoid, New York: State University of New York Press, 1990

- Gershom Scholem, "The Idea of the Golem", in Idem: On the Kabbalah and its Symbolism, Ralph Manheim (trans.) New York: Schocken Books 1965, 158–204
- Tzahi Weiss, Sefer Yesirah and its Contexts: Other Jewish Voices, Penn Press: Philadelphia, 2018

List of Figures

Before we begin

- To appreciate the discussions throughout this book, some familiarity with molecular biology is advised. Whenever a knowledge gap in biology is encountered, the biology library of Khan Academy might prove useful.
- Throughout the book, familiarity with the basics of differential calculus is assumed. A primer to calculus is available in the differential calculus course of *Khan Academy*.
- When discussing mathematical-electrical models, some familiarity with elementary electrical components is assumed. Preliminary knowledge in circuit analysis is available in the Circuit analysis course of *Khan Academy*. Furthermore, an excellent introduction to transistors is available in the semiconductor course of Khan Academy.
- Various parts in this book assume basic familiarity with artificial neural networks (feed-forward, convolutional, and recurrent networks). Readers are strongly encouraged to consult the work "Artificial neural networks for neuroscientists: A primer," written by *Guangyu Robert Yang* and colleagues [308] when needed.
- Throughout the book, various simulations and models are presented. The reader is encouraged to explore the book website where the code is shared. Detailed library dependencies and versions used are given on the book website. Please cite this book whenever the code is used. To appreciate and comprehend the code, familiarity with Python and NEURON is advised. The basics of Python are detailed at docs.python.org and the basics of NEURON at neuron.yale.edu. Examples of Python utilization in neuroscience are given in [216].
- The electrical simulations described throughout the book were performed using LT-Spice by Analog Systems. Learn about the

simulator at Analog.com. Some basic tutorials are available at learn.sparkfun.com

Glossary

ADALINE: ADAptive LInear NEuron

AER: Address Event Representation

ANN: Artificial Neural Network

API: Application Programming Interface

BC: Bipolar Cells

BCM: Bienenstock-Cooper-Munroe

BNN: Biological Neural Network

CMOS: Complementary Metal-Oxide-Semiconductor

CNN: Convolutional Neural Network

CPU: Central Processing Units

CR: Conditioned Response

CS: Conditioned Stimulus

DNA: Deoxyribonucleic Acid

DNN: Deep Neural Network

DOF: Degrees of Freedom

DSGC: Direction Selective Ganglion Cell

DSL: Domain-Specific Programming Language

DVS: Dynamic Vision Sensors

FLOPS: FLoating point OPerations per Second

fMRI: functional Magnetic Resonance Imaging

FPGA: Field-Programmable Gate Array

FPVA: Fully Programmable Valve Array

GABA: Gamma Aminobutyric Acid

GUI: Graphical User Interface

GP: General Purpose

GPS: Global Positioning System

GPU: Graphical Processing Units

HH: Hodgkin-Huxley

IC: Integrated Circuit

LGN: Lateral Geniculate Nucleus

LIF: Leaky Integrated-and-Fire

LMU: Legendre Memory Units

LSTM: Long Short-Term Memory

LTD: Long Term Depression

LTP: Long Term Potentiation

MOSFET: Metal–Oxide–Semiconductor Field-Effect Transistor

ND: Null Direction

NEF: Neural Engineering Framework

NGRAD: Neural Gradient Representation by Activity Differences

NLP: Natural Language Processing

NPM: Neural Programming Model

PD: Preferred Direction

PES: Prescribed Error Sensitivity

PID: Proportional, Integral, and Derivative

RAM: Random-Access Memory

ReLU: Rectified Linear Unit

RMSE: Root Mean Square Error

RNN: Recurrent Neural Network

RQC: Random Quantum Circuits

SPA: Semantic Pointer Architecture

SNN: Spiking Neural Network

STDP: Spike Timing Dependent Plasticity

SAC: Starburst Amacrine Cells

SiN: Silicon Neurons

SOPS: Synaptic Operations Per Second

SVD: Singular Value Decomposition

UR: Unconditioned Response

US: Unconditioned Stimulus

VLSI: Very Large-Scale Integration

I

Introduction and Overview

CHAPTER 1

Introducing the perspective of the scientist

Abstract

This chapter will introduce the scientist's perspective on neuromorphic engineering which explores the emergent properties of large-scale neuronal architectures. We will start by describing the neuron doctrine and move forward to brain modeling in both biological and computational systems. We will conclude by describing the benefits of using neuromorphic hardware to shed light on some of the brain's mysteries.

1.1 FROM THE NEURON DOCTRINE TO EMERGENT BEHAVIOR

Since *Alcmaeon of Croton* considered the brain as the cradle of the mind [53], scientists have tried to understand the existence of *qualia* – the reality of human subjective conscious experience, manifested by the brain [35]. About 2.5 millenniums later, confronted with the brain's scale and computational intractability and equipped with high-performance computers, neuroscientists are still trying to contemplate the brain's underlying mechanics. Some of them, inspired by *Richard Feynman's* immortal words: "*That which I cannot create, I do not understand,*" are trying to understand the brain by recreating it – in virtual or physical space [18].

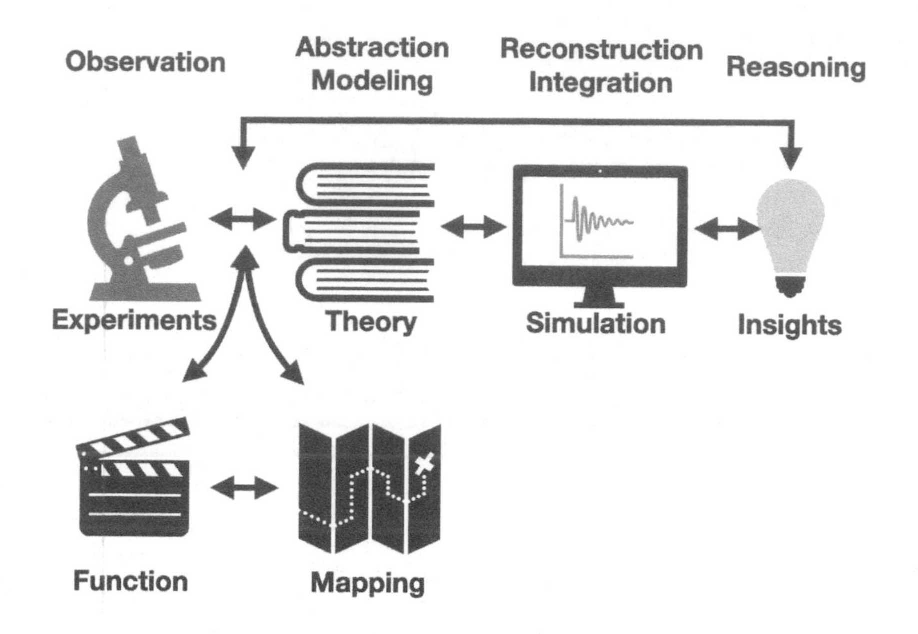

Figure 1.1 Neuro-scientific reasoning underlying the perspective of the scientist.

1.1.1 The unity and diversity of neurons

Neuro-scientific reasoning comprises intuitive thinking, experimental designs, measurement tools, and an immense amount of structured and unstructured data. While pure intuitive thinking might set the ground for experimental designs, measurements are used to construct theories and define relationships between function and structure. Data-driven simulations introduce experimental evidence from physical space into virtual space [91], aiming at fueling ideas and generating insights (**Figure 1.1**). This somewhat simplified scheme of thought underlies much of the perspective of the scientist.

More than a century ago, *Santiago Ramón y Cajal* described the brain as being comprised of individual interconnected cellular entities [180], later termed as *neurons*. The neuron was quickly crowned as the essential constituent of the nervous system, and later as the fundamental unit of perception [27], constituting the *neuron doctrine.*

The neuron has the canonical description of being coarsely composed of a *soma* (site of signals integration), *dendrites* (signal input pathways), and *axon* (signal output pathway). Neurons typically communicate with

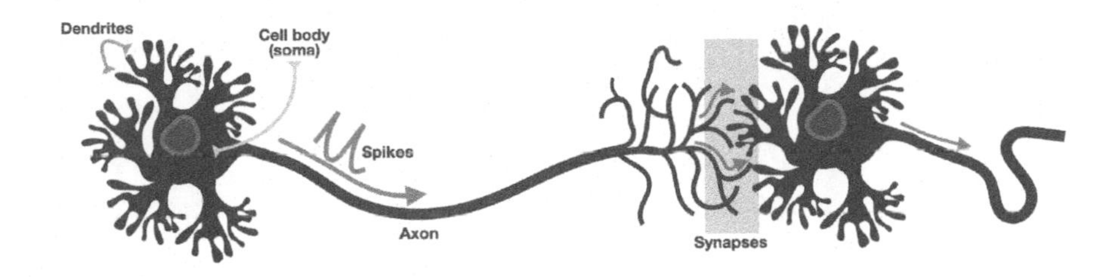

Figure 1.2 Canonical description of the neuron. The neuron is canonically comprised of a soma, dendrites, and an axon. Neurons typically communicate with spikes which propagate as impulses from the cell's soma through its axon to target neurons via synapses.

(many) other neurons with *spikes* – temporary changes in voltage which propagate as impulses from the cell's soma through its axon to target neurons via synapses (**Figure 1.2**). Neuronal communication can be therefore described as based on pseudo-binary, asynchronous messages. Spikes are referred to as pseudo-binary since, as we will see later, each conveys less than one bit of information.

What is the neuron's role? Arguing for the notion in which neurons are units of perception, a hypothetical neuron that responds only to a definite and significant stimulus, such as the image of one's grandmother, was conceptualized [110]. Studies demonstrated the existence of such neurons. For example, a neuron at the medial temporal lobe of a human's brain was found to fire as a response to all pictures of the actress Jennifer Aniston (when she was by herself). This "Jennifer Aniston neuron" did not fire (as many) spikes to other famous/non-famous faces, landmarks, animals, objects, or even to pictures of Jennifer Aniston with her ex, actor Brad Pitt [235]. Other studies demonstrated a neuron that solely responds to a video of the Simpsons (**Figure 1.3**). Furthermore, this "Simpsons" neuron also spiked when this person thought about the Simpsons (even before he was aware of it), without directly seeing the show [97], thus demonstrating the creative process by which reality is reconstructed in the brain.

Facing the homogeneous portrait given above, of having single-typed neurons which communicate with impulses stands a magnificent reality. Neurons are amazingly diverse, with hundreds of identified types. Neurons feature a great variety of morphologies and electrophysiological characteristics. Some neurons have 4.5-meter-long axons (giraffe primary

Figure 1.3 Neuron response to a preferred stimulus. Some neurons found to selectively respond to one preferred complex stimulus, such as the appearance of one specific figure in a particular situation.

afferent axon) while others have axons that are merely 0.0001-meter-long (granule cell). Spike generation is also nonhomogeneous across the nervous system, as some neurons generate spikes which travel as fast as the fastest train on earth (Shanghai Maglev train at $\approx$ 400 km/hour), and some generate spikes which travel slower than a sea turtle (Leatherback sea turtle at $\approx$ 2 km/hour). Some neurons are connected to hundreds of thousands of other neurons; others are connected to a mere few hundreds. Above this vast morphological and electrophysiological diversity, incomprehensible numbers are present. With $\approx 10^{10}$ neurons, $\approx 10^{13}$ synapses, and $\approx$ 72 km of fibers apparent in the human brain, the bare ambition to apprehend and understand this level of complexity is genuinely visionary.

Neurons are not only varied in morphology and physiology; they are tuned to prefer distinct stimuli. Through the development of electrophysiological recording methods, it was shown that a neuron has a *receptive field* in which it has a preferred stimulus: a specific feature of sensation which activates it. One of the earliest examples for a preferred stimulus is the directional selectivity observed in retinal cells. *David Hunter Hubel* and *Trosten Wiesel* showed that retinal ganglion cells selectively respond to movement in a distinct preferred direction [132]: a scientific pursuit for which they were awarded the Nobel Prize in 1981. They demonstrated that ganglion cells emit spikes in increasing frequency when an edge was moving in one specific direction across their receptive field (**Figure 1.4**).

1.1.2 Neural coding

Neurons mostly communicate with spikes. A sequence of spikes encode information using a diverse set of coding schemes [99]. In sensory

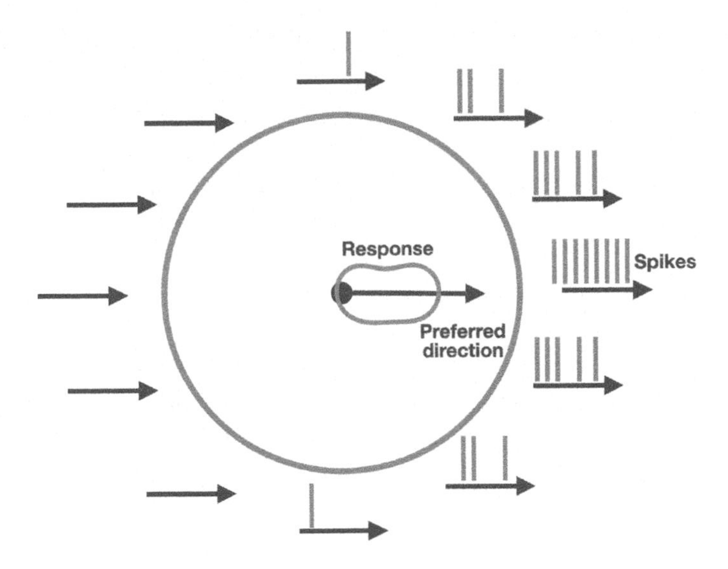

Figure 1.4 Ganglion cells selectively respond with an increased frequency of spikes for a certain visual modality or when an edge (sharp change in image's brightness) moves in one specific direction (preferred direction) across their receptive field. In this illustration, a neuron is stimulated with drifting bars of light. At 270 degrees, the bars move right to left and, at 90 degrees, the bars move from left to right. In this example, the cell's preferred direction is 90 degrees, for which it responds with a high spiking rate. The cell responds with less frequent spikes as the angular distance between the direction of the visual stimulus and its preferred direction increases.

neurons, where neuronal activity is driven by external, quantifiable inputs (e.g., sound and light intensities), spikes can be correlated to electrophysiological recordings, thus providing a glimpse to neural coding. The most straightforward approach to data encoding with spikes is *rate coding.*

With rate coding, information is encoded with the spiking rate. This coding scheme is rooted in experimentation. For example, the strength at which an innervated muscle is flexed was shown to (also) depend on its excitatory firing rate [88]; sound's intensity was related to spike firing rates in the neurons of the auditory tract [151]; and inferior colliculus neurons were shown to encode auditory azimuth angle, as their firing rate increases as a function of the distance from the sound origin to some perceptual reference anchor or fixation point [135].

A more elaborated computing scheme might emerge from rate coding at the population level. For example, *population rate coding* was suggested for sound localization and possibly for visual depth perception through the comparison of rate signals from the two sources (ears or eyes). Therefore, rate coding may be the brain's canonical description of location: encoding a perceived location through a population spike rate relative to some baseline.

However, identifying the spike rate with encoded information may be too one-dimensional. Even though rate code schemes are the easiest to develop systems for, they are not necessarily the most efficient in terms of resource utilization (e.g., energy) and performance. For example, decoding time might be too fast to allow for rate encoding. In the visual pathway, assuming $\approx$ 100 msec perception time, neurons activity propagate through 10 synaptic connections, thus permitting only 10 msec for processing time per synapse. Accounting for spike generation times, it allows for only one or two spikes to be generated. This is not enough spikes to enable accurate encoding. Such constraints can be alleviated with *temporal coding*, or "time to spike" encoding [276]. Interestingly, several works demonstrated that a combined rate and temporal codes could be used for a spatio-temporal coding with interesting features [8]. Coding schemes are summarized in (**Figure 1.5**).

Neural coding is far from being entirely deciphered, particularly in cases where nonlinear feedback between neurons is used and when the non-deterministic nature of spike generation is considered. These intractable characteristics arise from the fact that a synapse will often not respond the same for similar inputs, as previous stimuli modulate its state. This adaptive behavior is often referred to as *synaptic depression/adaptation*. When a synapse adapts to a stimulus, its response becomes inversely proportional to the stimulus' magnitude.

Neural coding, therefore, remains a mystery. Given sensory inputs, various combinations of neural networks respond, sending feedback to one another, giving rise to intractable spikes. Some of these spikes affect the generation of other spikes, and others might not generate a response at all. Those information-carrying spikes continue to propagate through massively parallel networks, without having central guidance nor a synchronizing "master" clock. Amazingly, this orchestra of spikes gives rise to a (hopefully) stable manifestation of coherent consciousness.

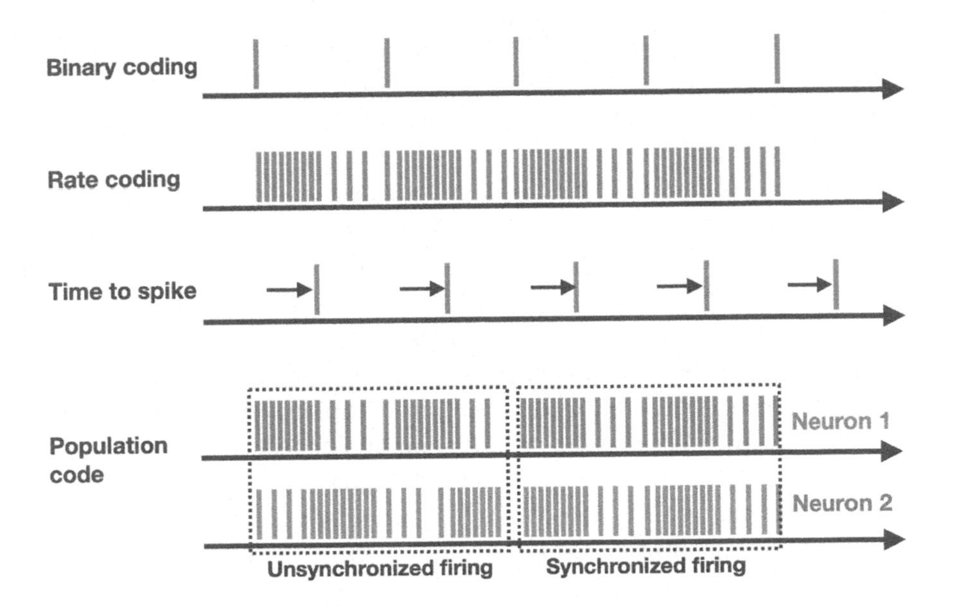

Figure 1.5 Neurons encode information with spikes. Encoding can be based on binary coding (responding when something happens), rate coding (information as a function of spiking rate), time-to-spike (information as a function of the exact time at which a neuron fired), and with population coding (information as a dynamical pattern of firing).

1.1.3 Networks and emergent behavior

In contrast to the neuron doctrine, network-focused models suggest that a function can emerge from the joint activation patterns of interconnected ensembles of neurons. These neuron ensembles can generate emergent functional states which cannot be observed by studying the single entities they comprise.

Emergent states are not unique to the brain. They can be found in bird flocks, where each bird contributes to the formation of one flock in which all birds fly with the same velocity, without central guidance [118]. Like the brain, just as each neuron knows nothing about the conscious experience it helps to generate, each bird in the flock knows nothing about the flock's behavior. Although both the bird and the neuron follow simple rules, the population's aggregated effect becomes an emergent property. Another prominent example of emergent behavior is the Game of Life, a cellular automaton devised by *John Horton Conway* in 1970. Conway wanted to show how cellular automata can produce exciting and surprisingly complex behaviors. In the Game of Life, each square cell residing in a two-dimensional grid follows a simple set of rules: any live cell (white square) with two or three live neighbors survives; any dead cell (black square) with three live neighbors becomes a live cell; all other live cells die in the next generation, and all other dead cells stay dead. The initial configuration of the grid (which cell is alive and which one is dead) determines its fate. While some arrangements last for a long time before demising, others go on indefinitely. Furthermore, it was shown that by carefully setting the initial conditions of the game, one could generate sliding text and images, compute logical functions, and create self-replicating machines. This complicated behavior emerges with no central coordination, where none of the cells are aware of what they help to develop [5]. Numerous such initial configurations are developed routinely; one of them is the crawling "lobster." By carefully setting the game configuration, the emerged cellular automaton looks like an animated form moves through the lattice. A running example of a crawling "lobster" is shown in **Figure 1.6** (the Lobster was voted the Pattern of the Year 2011 on ConwayLife.com).

To what extent can we consider cognition to be an emergent property of the nervous system? Can we expect intelligence to emerge from randomly connected neurons, or should some guided order be introduced? In his article from back in 1988, *William Daniel Hillis* wondered [122]:

> Emergence offers a way to believe in physical causality while simultaneously maintaining the impossibility of a reductionist explanation of thought. For those who fear mechanistic explanations of the human mind, our ignorance of how local interactions produce emergent behavior offers a reassuring fog in which to hide the soul.

The "reassuring fog" given to us by the idea of emergence is something many neuroscientists still find comforting. In the next sections, we will delve into some of the ways emergence is sought after. However, before going down the rabbit hole, we will discuss neuronal abstractions.

1.1.4 Neuronal abstractions

To what level of detail should the brain be investigated? Should we concentrate on the molecular level which gives rise to neurons and synapses, or should we focus on the level of networks, maps, systems, or beings?

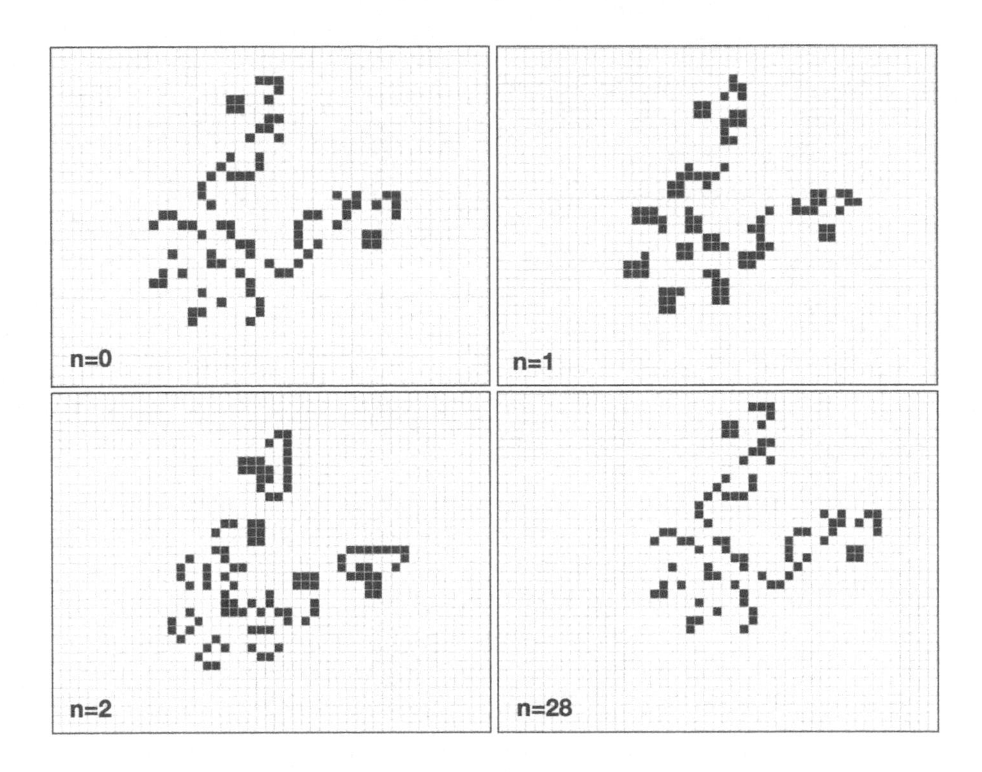

Figure 1.6 Animation of a crawling lobster (four frames taken at different steps of the simulation) created with the Game of Life cellular automaton, demonstrating emergent behavior.

The level of abstraction used to study the brain remains a crucial question as complex behavior could be inferred from data obtained across multiple spatial and temporal scales (angstroms to centimeters, milliseconds to years) (**Figure 1.7**).

Let's take the process of decision-making as a guiding example. According to the standard economic accounts of human rationality, a rational agent would assess the probability and potential reward for each available action before making a decision and decides with maximal gain. Can we take this behavioral observation and be satisfied with its explanatory power? Being biological beings, would a biological explanation be more satisfying?

One can easily argue that the brain treats decision-making as an optimization problem and is driven to do so by virtue of evolutionary pressure. This ecological view argues for the evolutionary benefit of being able to optimize the speed and accuracy of decisions and to maximize

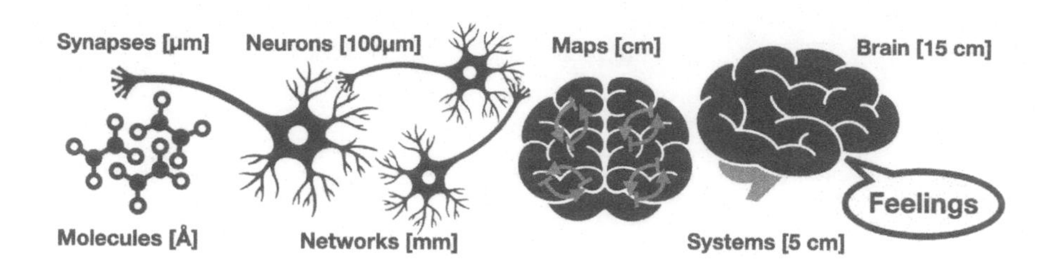

Figure 1.7 A brain model can relate to various abstractions of the brain: from the molecular level (10^{-10} m) to synapses (10^{-6} m), neurons (10^{-4} m), networks (10^{-3} m), maps (10^{-2} m), systems ($5 \cdot 10^{-2}$ m), and the whole brain ($1.5 \cdot 10^{-1}$ m).

the rate of receiving rewards for the right choices [43]. Is this explanation sufficient? Considering that decision-making is a cognitive process, shouldn't a neurophysiological explanation be considered a better one?

A seminal work by *Michael Platt* and *Paul Glimcher* linked simple decision-making to neuronal activity in the parietal cortex in monkeys [229]. They showed that neurons could encode the relative amount of gain that the monkey can expect from making a particular decision. More recently, a choice between alternatives was shown to incorporate intricate neuronal activities in the basal ganglia; in particular, that one of the basal nuclei, the subthalamic nucleus, has a role in modulating the decision process proportionally to the conflict between evidence for other alternatives [43]. Interestingly, it was shown that economic optimization-driven theories could generate experimental predictions for both neurobiology and behavior.

Human beings, however, are proudly known for being *irrational.* For example, humans often rely on intuition or their emotional responses for guidance and can therefore be manipulated (e.g., through the "framing effect"). This, again, was found to have roots in neurobiology. It was found that framing and biases are associated with amygdala activity, suggesting a key role for an emotional system in mediating decision biases [71].

Decision-making can, therefore, be formulated on different levels of abstraction, from economical to ecological to neuro-physiological. It ranges from detailed models of neural circuits to abstract psychological models of behavior. The most satisfactory explanation would take an integrated approach to neuroscience. Ideally, it requires multi-scale neural data – from molecular regulations and the dynamics of individual synapses to pattern processing in neural networks, the orchestrated function of brain maps, and the formation of neuronal systems. Generating the vast amount of data required to generate such integrated models is a central theme in many brain-tailored research consortiums, which will be briefly discussed in the next section.

1.1.5 Data-driven neuroscience

The level of expertise and the amount of data required for detailed neuronal models is an important bottleneck. Collaborative brain science is, therefore, emerging as an essential endeavor. To support collaborative research, developing accepted and well-defined models to allow data sharing is requisite [66, 284].

One of the most ambitious brain-focused research initiatives is the European Union's Human Brain Project (HBP) which aims at bridging the scales between the neuron and system levels in the brain, integrating electric, magnetic, and optical signals measured at the population and system levels [81]. To support such vast research, entire frameworks for neuroinformatics were constructed. Neuroinformatics lay at the core of HBP and orchestrated by COLLAB, a collaborative cloud-based system developed within HBP's neuroinformatics platform. COLLAB fuels the project's other platforms (brain simulation, neurorobotics, medical informatics, and neuromorphic computing) with immense upstream and downstream data flows. It is powered by HBP's high-performance analytics and computing platform, enabling massive data archiving and distribution of virtual machines to collaborators, empowering them with high-end supercomputing capabilities for simulation and data analytics. COLLAB's mission is not a trivial one: it must be interfaced with heterogeneous data types and ontologies to manage metadata storage and provide a query system with which rodent and human brain atlases can be constructed and populated using different data modalities (anatomy, physiology). Moreover, COLLAB should link its data with foreign maps, databases, and atlases. Inspired by HBP, another scientific endeavor termed "BRAIN" was initiated in the United States, "aimed at revolutionizing our understanding of the human brain" [139] and like other "big-science" initiatives, to "empower individual labs by providing open-access databases" (e.g., The *Open Source Brain* project [105]).

It is often argued that simulating neuronal activity is not sufficient for brain modeling. Realistic models should also encompass the reconstruction of the cellular components in which biochemical reactions occur, including astrocytes (supporting cells) and the blood vessels that deliver oxygen, glucose, and other nutrients to the neurons, affecting their dynamic [59]. Some data-driven initiatives take into account heterogeneous biological data, aiming at an even higher level of integration. For example, the National Institutes of Health-funded Human Connectome Project (HCP) aims to characterize human brain connectivity and functions. In this project, a colossal amount of data are gathered from many hundreds of patients with 3D fMRI machines, Electroencephalogram, and Magnetoencephalography. Full-genome sequencing from all subjects are performed and behavioral measures in different domains (cognition, emotion, perception, and motor function) are recorded [288]. Other governmentally-funded integrated neuroscience programs include "Brain Canada" [145] and the "China Brain Project" [230]. All aforementioned acknowledge the importance of an integrated data-driven approach while emphasizing that establishing standardized data collection and processing and mechanisms for data sharing and credit allocation are fundamental to their project's success.

Data-driven neuroscience emphasizes that from the scientist's perspective, to devise models with high explanatory power, truly multidisciplinary expertise is required.

1.2 BRAIN MODELING

Numerous main modeling paradigms were developed to provide a model of the brain with high explanatory power. Here we will briefly discuss two of them. First, brain modeling in biological systems where tissue engineering and genetic engineering are employed to create a system in which various nervous system aspects can be investigated will be discussed. Afterward, the computational modeling of the brain in which mathematical descriptions of neuronal behavior are implemented on a large scale will be described. Large-scale neuronal modeling cannot be efficiently executed on traditional computing hardware, giving rise to the idea of utilizing neuromorphic hardware for a more efficient realization. This will be the topic for the last section.

1.2.1 Brain modeling in biological systems

As the sheer complexity of the brain is staggering, artificially organizing neurons' complexity and their connections seems impractical. One way to reconstruct parts of the brain and study it in-vitro is by studying *brain organoids*. Brain organoids are self-organized stem cell-derived living human neurons, cultured in bioreactors within 3D scaffolds, first reported by *Medaline Lancaster* in 2013 [163]. Medaline demonstrated the formation of self-assembled neuronal mosaics in which the cultured living cells reorganized in space to generate different brain regions. Later developments included region-specific organoids which were used to study cortical folding and brain development disorders [164]. Brain organoids were even implanted into a living rodent host with which the organoids established functional synapses with the mouse neurons, exhibiting synchronized neural activity [189] (**Figure 1.8**). The organoids were vascularized by the mouse brain and survived for a prolonged time. Biological tissues do not function in isolation; they are connected, comprising intricate functional sub-systems and systems. To better understand system-level interaction, brain organoids were used in conjunction with other organoids, creating assembloids. For example, *Sergiu Pasca* and colleagues assembled region-specific organoids of the human striatum with cerebral cortical organoids to form cortico-striatal assembloids in which they demonstrated functional synaptic connections. Interestingly, the researchers derived cortico-striatal assembloids from patients with a neurodevelopmental disorder with which they were able to monitor disease-associated defects, thus demonstrating the potential of using assembloids to investigate the development and functional assembly of brain connectivity [211].

Using brain organoids as a model for the brain can be useful for some particular aspects of brain science. However, they entail a few fundamental constraints. In comparison to a real human brain, brain organoids are simple in their structure and connectivity scheme. They are not fully controllable and, most importantly, they are not programmable, hard to maintain, and difficult to utilize to study large scale dynamics. Some ethical concerns were raised regarding the development of brain organoids [166], concerning their potential having some emerged conscious experience. However, the consensus is that brain organoids currently lack the sensory inputs and the complexity needed for a chance of having such emerged characteristics.

Another approach to empirically study the uniqueness of the human brain is to study its underlying genetics. In a study led by *Bing Su*, rhesus monkeys were genetically engineered to carry human copies of the MCPH1 gene, an imperative gene for brain development in humans [257]. The monkeys experienced a human-like delayed neuronal

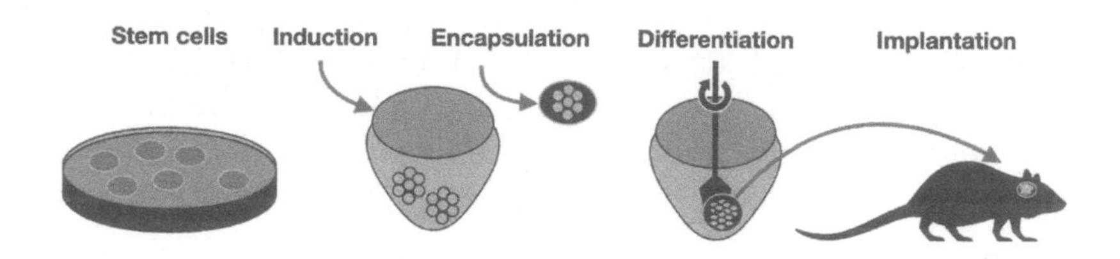

Figure 1.8 Brain modeling with organoids. Brain organoids are self-assembled three-dimensional cellular assemblies. Stem cells dissociate from mouse embryonic cells, undergo neural induction in culture, encapsulate within a gelatinous protein mixture, and transfer from stationary culture to a spinning bioreactor containing differentiation media. Finally, the organoids are implanted into a mouse brain.

maturation and axon myelination and, remarkably, demonstrated better short-term memory and shorter reaction time than wild-type controls. While remarkably interesting, this physical reconstruction's explanatory power to better understand human brain traits is relatively limited, as the effect of genetic manipulation, even when accurate and controllable, can have untraceable systematic effects.

To conclude, brain modeling in biological systems holds great promise in many respects. However, to study the brain's intricacies, we need a model in which all variables can be precisely isolated and controlled. The details, which have to be modulated and examined, are vast and impossible to be experimentally addressed. For example, neurons have been shown to have different activity patterns regarding their electrophysiological characteristics, proteins and neuropeptides expression profile, and morphological features. They are connected in vast networks with synapses strategically positioned at particular locations on particular morphological sections. Experimental models cannot offer the tools with which such convoluted organization can be efficiently addressed. To do that, a computational model is essential.

1.2.2 Computational brain modeling

Most large-scale biologically plausible models of the brain concentrate on the *cerebral cortex*. The cerebral cortex is a sheet of few millimeters thick of tissue at the brain's surface. It is organized horizontally in six layers and vertically in millions of columns, hundreds of microns in diameter, constituting the *cerebral column*. Within the cerebral column, a limited lateral spread of information is apparent, and most activity is intra-laminar. Cerebral columns have neurons with corresponding characteristics, suggesting that the columns constitute functional-structural entities. Columns are interconnected to create networks of cortical areas (millimeters in size) which have high cognitive functions, such as representing sensed data from a particular area of the body. Two main principles can guide the construction of computational models:

1. Neuronal abstractions. One of the great challenges of brain modeling is the difficulty in assessing the appropriate level of neuronal abstraction at which the model should aim. Neuronal abstractions were described in Section 1.1.4.

2. Modeling strategy. Most highly detailed computational modeling approaches can be referred to as *bottom-up modeling*: building the brain as accurately as possible from the very fine details of the brain's building blocks, slowly constructing whole neuronal systems. In this approach, we start with relatively simple entities, gradually developing the relationships among them. A different approach is *top-down modeling*. In a top-down approach, the computational model is constrained to reproduce an observed cognitive capability. Thus, it reverses the direction of exploration and promotes evaluation of neuronal architecture from which a particular behavior can emerge [86] (**Figure 1.9**).

What constitutes a brain model is, therefore, a matter of perspective, as well as a matter of the question at hand.

1.2.2.1 Bottom-up modeling

Bottom-up modeling is a central neuronal modeling paradigm in which models are described with molecules and mathematical processes, accounting for networks with high explanatory power.

In 2009, *IBM Research* released a technical report in which they presented a cortical simulation of 10^9 neurons and 10^{13} synapses. IBM celebrated this achievement with the title "The cat is out of the bag," referring to the estimated size of a cat brain [15]. In 2012, IBM released another report with the title "10^{14}," declaring a simulation of 10^{10} neurons and 10^{14} synapses on IBM's BlueGene supercomputer (referring to the estimated number of synapses in the human brain) [306].

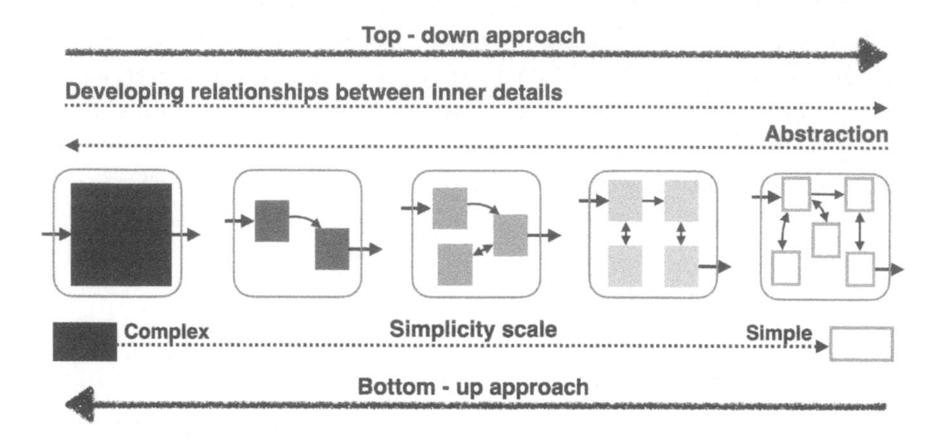

Figure 1.9 In a bottom-up approach, we start with simple "bright" entities, slowly constructing the whole system by progressively developing relationships between them. As we move up the abstraction ladder, interconnected entities are encapsulated into "darker" abstracted entities. In a top-down approach, we start with the system itself, slowly breaking it apart into simpler inter-connected entities.

These simulations, however, have mathematically abstracted away cellular morphology, biophysical properties, and synapse dynamics and therefore, while representing a remarkable engineering achievement (for which the research team was awarded the *ACM Gordon Bell Prize*), are far from being plausible computational models of biological brains. This research is an example of using bottom-up modeling, where the fundamental entities are simple mathematical descriptions.

One of the most influential advocates for the importance of biologically plausible models is *Henry Makram*, the director of the *Blue Brain Project.* In an open letter to IBM, Makram portrayed these simplified models as "trivial," for which awarding the Bell Prize "is beyond belief" [7]. One might argue that simplistic models cannot demonstrate the "quantum leaps in the 'quality' of intelligence" which, according to Makram, are essential to biological intelligence. The Blue Brain Project aspires to represent the other extreme of the debate, where highly detailed biological models are simulated on computers with "petaFLOPS speeds" – computers capable of performing 10^{15} FLoating point OPerations per Second (FLOPS). The project aims to use these models to provide insights into the "emergence of biological intelligence" [190]. In a fascinating study published by the Blue Brain Project consortium, a biologically plausible model of the cortical column was presented [192]. In this study, $\approx$31,000 neurons, featuring hundreds of morphological and electrical types and $37 \cdot 10^6$ synapses, were simulated within a volume of 0.01 mm^3.

Both detailed and abstracted modeling paradigms have reproduced empirically observable spike patterns, activity waves, and rhythms. However, hoping for some sort of intelligence or interesting behaviors to emerge from them is still considered by many as wishful thinking [86]. Nevertheless, detailed large-scale neuronal simulations pose to take our understanding of emergent activities a step forward. For example, by comparing two cortical microcircuit models, one with simple morphological structure and the other with higher-order design arising from morphological diversity within neuronal types, differences in the emergent activity was observed. Increasing the morphological detail level leads to more heterogeneous distributions of synapses and channels and more subtle changes in neuronal firing patterns. It also increases the number of cliques and contributes to small-world network topology (While most neurons are not connected, most neurons can be reached from every other neuron by a small number of intermediate neurons.) Thus, the higher-order structure imposed by morphological diversity within neuronal types impacts emergent cortical activity [220].

1.2.2.2 Top-down modeling

For many, top-down modeling makes much better sense. For example, if we were trying to understand the way a spider moves through space by seamlessly configuring his body postures, exploring the detailed composition of its tarsal claws, tarsus, metatarsus, tibia, patella, and coxa (that comprises each of his six legs), this might not be the most efficient way do it. A more efficient approach would be to simplify the spider's legs into mechanical assemblies, exploring its behavior and designing a constrained mechanical model. Similarly, constraining a brain model to known cognitive characteristics might be a more efficient modeling paradigm.

Top-down modeling might also be a better reflection of biology. For example, the traditional bottom-up approach to vision processing, being initiated in the retina and propagating forward through processing modules with increasing complexity, is inadequate. According to higher-order cognition, it seems like a top-bottom approach affects the operations performed by cortical neurons (re-entrant processing) [101]. With top-down modeling, neurons tune their responses to stimuli by changing the correlations' structure over neuronal ensembles. This idea is at the foundations of the *Semantic Pointer Architecture* (SPA).

SPA is one prominent example for top-down modeling, proposed by *Chris Eliasmith.* SPA is a framework with which neuro-anatomically and physiologically constrained neural models can be optimized to produce unified cognitive neural networks. It revolves around the concept of semantic pointers, with which semantic content (e.g., the color blue, the rectangular shape) is represented with neural activity. SPA is described in [83]. SPA provides the framework with which semantics, syntax, control, learning, and memory can be represented with lower-level entities. In a seminal work, Eliasmith used SPA to construct *Spaun*, a $2.5 \cdot 10^6$ neuron model of the brain, capable of responding to visual stimuli by manipulating a virtual arm [87]. Spaun was shown to execute a diverse set of tasks, including digit recognition, tracing from memory, serial working memory, question answering, addition by counting, and symbolic pattern completion. Spaun 2.0 was reported to feature twice the number of neurons and was shown to perform cognitive tasks ranging from digit recognition to inductive reasoning [57].

The SPA's explanatory power is impressive as it allows relating functional behaviors to the level of spiking neurons. This was accomplished by adjoining SPA to a bottom-up modeling framework – the *Neural Engineering Framework* (NEF), also proposed by Chris Eliasmith [84]. Eliasmith devised a set of principles, which underlie NEF, enabling the optimization of a spiking neural network to realize the transformation and dynamics of the desired behavior [84]. While the SPA details are outside the scope of this book, NEF will be described in Chapter 12. Interestingly, in an attempt to increase the explanatory power of Spaun, it was shown to be operational when its underlying spiking behavior is based on biologically plausible (conductance-based) neuronal models [85]. This biologically plausible cognitive model, named "BioSpaun," was used to investigate pharmaceutical intervention. Notably, it was used to estimate the effects of applying TTX, a sodium channel-blocking drug on cognitive performance. The model showed promising results in its ability to replicate human reaction times (in terms of reaction time and task completion) in response to TTX-induced failure.

Both modeling paradigms require an immense amount of computational resources. To computationally evaluate intricate models, neuromorphic computing offers unique advantages. This will be the topic of the next section.

1.2.3 Brain modeling with neuromorphic hardware

Large-scale neuronal modeling requires an immense amount of computing power. For example, the cat cerebral cortex is estimated at 10^9 neurons and 10^{13} synapses. An efficient representation of 16 bytes/synapse in memory requires 160 terabytes of main memory to allow for near real-time simulation. Assuming a neuronal firing rate of 1 Hz, a clock-driven simulator would have to support a total of 10^{13} messages/second, requiring a significant amount of ingenuity in designing the communication architecture. Therefore, to take the "cat out of the bag," IBM used a supercomputer with 147,456 Central Processing Units (CPUs) and 144 TB of total memory. IBM human-scale simulation of the brain was implemented on a 96 (racks) $\times$1,024 (nodes) $\times$16 processing cores computer with a total of 1.6 petabytes of memory, covering an area of about 3,000 square feet (280 m^2). This petascale computing power required an immense amount of energy. Powering the demand for $\approx$ 20 peta FLOPS (PFLOPS) on the BlueGene requires $\approx$ 8 megawatts of power. And this is with the computer which consistently leads the *Green500* rankings of the most powerful and energy-efficient supercomputer available [116]. All this computational power is dedicated to simulating a dramatically abstracted brain model at the speed of x1,542 slower than real-time. What about simulations of highly detailed models, such as the one conducted by the Blue Brain consortium? Their model (consisting of "only" 31,000 of biologically plausible neurons) was also executed on IBM's BlueGene supercomputers. Their design was implemented in the Swiss National Supercomputing Center on a computing system, ranked the 100th most powerful in 2015 [192]. What about Spaun? The Spaun model was executed on a 24 GB RAM-powered computer. It needed 2.5 hours to simulate 1 second. Fortunately, later developments in both hardware and algorithms allowed for efficient execution of NEF/SPA over Graphical Processing Units (GPUs) [107]. Exploiting the thousands of computational cores embedded within a modern GPU for Spaun simulation yielded a 390-folds speed-up. While Spaun can now be efficiently simulated in terms of execution time, power consumption remains high as simulating Spaun in real-time would require $\approx$ 6 kilowatts of GPU power. Considering that the human brain is estimated to consume only

20 watts, the computing architecture on which large-scale neuronal simulations take place is not adequate.

Neuromorphic hardware can support the simulation of neural networks with increasing size and biological plausibility, thus providing insights into the intricacy and the emerged properties of the brain, theoretically, in real-time, and with low energy requirements. We will discuss neuromorphic architectures later in Chapter 8. One of the earliest designs for a neuromorphic computer was the NeuroGrid which was proposed by *Kwabena Boahen* in 2014 [33]. The NeuroGrid can execute a neural model with 10^6 neurons and $8 \cdot 10^9$ synapses while consuming merely 3.6 watts altogether, allowing for the simulation of cortical models in real-time. Simulating Spaun on the NeuroGrid can potentially consume less than 1 watt, enabling it to run for 24 hours on a single 1.5 V AA battery [42]. An NEF/SPA-tailored hardware design is the BrainDrop [217] – the successor of the NeuroGrid, demonstrating the hardware-software co-design paradigm (see also the case study in Section 8.3).

To conclude, from the scientist's perspective, one of the appealing key points in neuromorphic engineering is the potential to efficiently execute large scale neuronal simulations – close to real-time and with modest energy requirements.

1.3 GLOSSARY

Bottom-up and top-down modeling: In bottom-up brain modeling, large-scale networks are defined using its elementary building blocks. In top-down brain modeling, low-level details are constrained to higher-level observations.

Networks and emergent behavior: A function can emerge from the joint activation patterns in neural networks. Emergent behavior cannot be observed by studying single entities.

Neural coding, rate coding: A spike train encodes information using a coding scheme. In a rate coding scheme, data is encoded by the spiking rate.

Neuron: A nerve cell, comprised of a soma (site of signals integration), dendrites (signal input pathways), and axon (signal output pathway). Neurons typically communicate with other neurons with spikes.

Neuronal abstractions: The brain can be investigated from different abstraction levels ranging from the molecular level, neurons, and synapses to networks, maps, systems, and beings.

The neuron doctrine: The neuron is the essential constituent of the nervous system and the fundamental unit of perception.

1.4 FURTHER READING

- **Section 1.1.**
  - Learn more about the roles of single neurons in perception in [27], a classic paper by *Horace Barlow.* This paper was later updated by Barlow and colleagues in 2009 [26].
  - In [311], *Rafael Yuste* describes one of the most important paradigm shifts in neuroscience: shifting from the investigation of single neurons to studying the emergent properties of neural networks.
  - *Torsten Wiesel* tells us the story of the Nobel Prize-winning discovery of how the brain processes information through the visual cortex, through receptive fields and preferred stimuli in iBiology Science Stories YouTube channel.
  - In [235], *Itzhak Fried* and colleagues report on the observation of having single neurons responding to one particular complex stimulus, such as a face of a celebrity.
  - In [242], *Paul Rendell* discusses the construction of a Turing Machine from Conway's Game of Life, demonstrating a behavior emerged from simple sparsely interconnected entities which follow a limited set of rules.
- **Section 1.2.**
  - In [163], *Madeline Lancaster* presents her seminal work on the development of brain organoids and in [164], its expansion into area-specific organoids (forebrain). In [189], *Abed AlFatah Mansour* discusses implantable vascularized brain organoids, and in [166], *Andrea Lavazza* discusses ethical issues of using these organoids. Assembloids are succinctly discussed in [224] and described with further details in [194].
  - In [257], *Shi Lei* and colleagues describe their attempt to genetically engineer "smarter" rhesus monkeys.

- In [119], *Suzana Houzel* outlines the staggering numbers which characterize the brain, along with an interesting discussion about the development traits in the evolution of the human brain.
- A review describing the history of large scale simulations of neuronal dynamic is given in [91].
- In [115], the canonical structure of the cortical column is described, along with some early guidelines for its simulation.
- In [14], *Rajagopal Ananthanarayanan* and colleagues dive into IBM's attempts to simulate large-scale neural networks. In [15] and [306] IBM's attempts to simulate neural networks of the scales of a cat and human brain, respectively, are described. IBM's abstracted simulations initiated a debate with *Henry Markram*, termed *"The cat-brain fever"* described in [7] by *Sally Adee.*
- In [190], *Henry Markram* describes the Blue Brain Project, and in [192] one of the project's flagship achievements – a detailed simulation of a cerebral column – is described. Criticism regarding the Blue Brain Project concerning its declared set of goals and management are described in [274].
- In [86], *Chris Eliasmith* and colleagues describe the pros and cons of the different approaches in large-scale neuronal modeling. They discuss the required level of biological plausibility and compare bottom-up and top-down approaches for modeling. In [87], they present their top-down modeled functional brain model which they named Spaun.
- A description of the explanatory power of top-down modeling in biology is described by *Giovanni Pezzulo* and *Michael Levin* in [227].
- Detailed power estimation of IBM's BlueGene supercomputer is given in [116], and optimization of NEF's execution time with Graphical Processing Units is given in [107].
- In [33], *Ben Varkey Benjamin* provides a detailed description of the NeuroGrid, a neuromorphic circuit developed by *Kwabena Boahen* and colleagues. In [42], Kwabena Boahen describes the NeuroGrid from a higher standing point while describing the implementation of NEF on the board.

CHAPTER 2

Introducing the perspective of the computer architect

Abstract

Acquiring the capability to spatially and temporally manipulate electrons within integrated circuits which typically comprise billions of transistors has changed the course of human society and practically defined the digital age. However, advancements in the fabrication of integrated circuits, which for two decades were driven by Dennard Scaling and Moore's law, are slowly declining due to quantum and power constraints. One promising direction for architectural chip design which aims at providing computational resources with low energy and high performance is neuromorphic engineering. This chapter will introduce the computer architect's perspective on neuromorphic engineering which aims to provide advanced computational resources with low energy and high efficiency. We will discuss some of the limitations in designing integrated circuits, understand the architectural rationale of neuromorphic designs, and introduce some of the prominent neuromorphic frameworks currently available.

2.1 LIMITS OF INTEGRATED CIRCUITS

At the foundations of the digital age stand an astonishing number of transistors. Transistors can be used to define logical gates with which memory chips, Arithmetic Logic Units (ALUs), and control circuits can be designed. These can be assembled to create CPUs, featuring billions of transistors. The story of the development of the CPU is fascinating as it encompasses the rise and fall of titanic companies, entangled by some of the most complicated structures ever designed. In their outstanding book "Computer Architecture: A Qualitative Approach," *John Hennessy* and *David Patterson* outline the incredible advancements in computer design over the past century [117]. By comparing computer performance relative to the VAX 11/780 (a minicomputer developed by DEC back in 1970) through the evaluation of numerical computation benchmarks, the increased level of performance of digital computers can be traced back.

For almost two decades, starting from 1986, we have witnessed an exponential growth in computing power at an annual rate of 52% (doubling performance every two years). For example, the Intel Xeon EE 3.2 GHz CPU developed during early 2000 is 6,043 times more powerful than the VAX11/780, developed only 20 years earlier. This trend was coined as *Moore's law*, named after *Gordon Moore*, the CEO and co-founder of Intel (**Figure 2.1**).

However, in the past two decades, the improvement pace of performance advancements is slowing down. Since 2015, growth in computing power is estimated at an annual rate of 3.5% (doubling performance every 20 years) (**Figure 2.1**), thus announcing the demise of Moore's law. From an economic point of view, the current stagnation of performance advancements is reflected by the cost of transistors. For the decade stretching from 2002 to 2012, we have seen increased economic growth in terms of the number of transistors which can be purchased for a dollar. Along with the improvement in transistor fabrication, until 2012, a dollar could have bought more transistors every year. For example, in 2002, a dollar could buy two million transistors (with 180nm fabrication); by 2012, the same dollar could buy 20 million transistors (with 28nm fabrication). This rapid economical growth halted in 2012. Although we can buy better computers (mainly by adding more cores to our computing arsenal or improving heat dissipation), we would have to pay significantly more to get them (**Figure 2.2**).

To better understand the roots for the demise of Moore's law as well as to appreciate the emergence of new computing paradigms, we will explore the following approaches which were taken in an attempt to increase computer performance:

- Packing more transistors in a chip
- Using a higher clock rate for processing speed

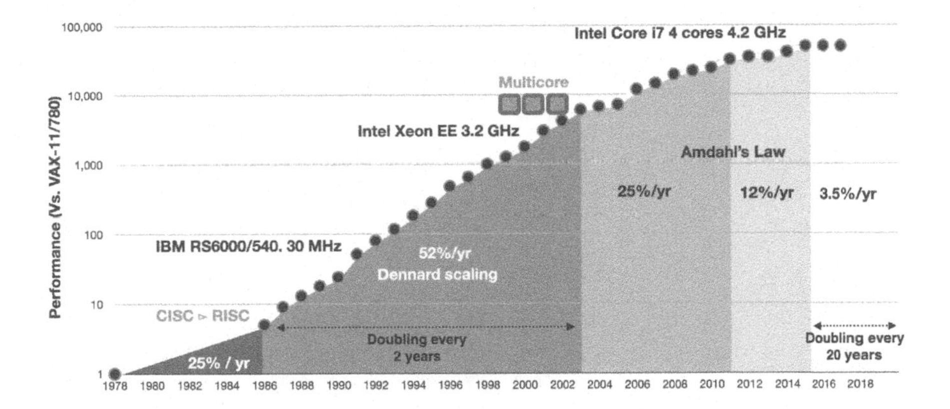

Figure 2.1 Computing performance calculated year-to-year from 1978 to 2018, compared to the VAX 11/780 computer, using standard numerical computation benchmarks. From 1986 to 2002, we have witnessed exponential computing power growth at an annual rate of 52%. Since 2003, single processors' performance slowed down, growing at the rate of only 25% a year. From 2011 to 2015, the annual improvement was 12%, and since 2015, with the demise of Moore's law, the recorded improvement rate slowed down to 3.5%. Overall, from doubling performance every two years during the 1980s, we are currently doubling performance every 20 years. Data from [117].

- Moving on to distributed computing with multi-core designs
- Changing computing paradigm

2.1.1 Transistor density

The first approach is concentrated around diminishing the transistor size. However, it seems that we are currently about to hit a fundamental wall. One of the driving powers of Moore's law is *Dennard Scaling*. This scaling law, observed by *Robert Dennard*, argued that as transistor density doubles, chip performance improves while its power density remains the same. It means that by making our transistors smaller, we get improved performance without having to pay for it with energy. Dennard understood that a smaller transistor would need less current and a smaller voltage for its operation, leading to reduced energy consumption. However, integrated circuits are limited regarding the extent of the

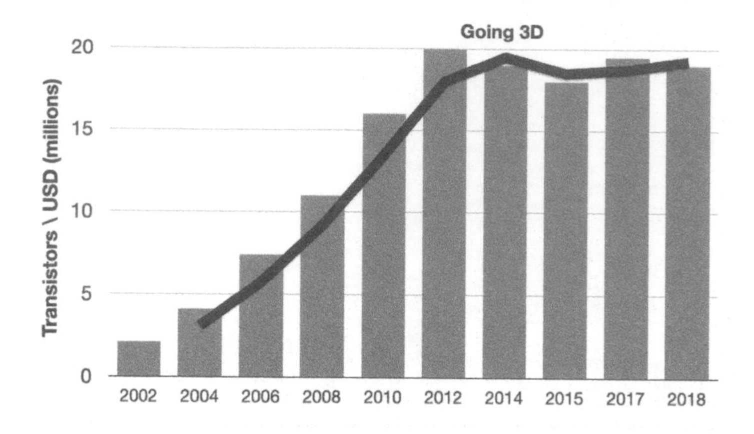

Figure 2.2 Trend in transistor cost. For more than a decade, from 2002 to 2012, we have witnessed tremendous economic growth, where with each passing year, transistor cost dropped. However, since 2014, this growth has stagnated. In 2012, for the first time in computing history, a transistor's cost went up due to the introduction of more expensive 3D fabrication processes. Data from [42].

current and voltage drop they can support, bringing Dennard Scaling to an end in 2004, 30 years after it was observed.

Transistors are essentially nano-scale switches for electrons. In a planar transistor design, electrons are driven from the transistor's drain to its source and controlled by the voltage applied to its gate. With a 30 nm design, six 5 nm wide lanes are available for the travel of six electrons across the transistor. As electrons travel over the transistor's silicon-dioxide interface, they could be trapped by the stochastically and predictably present *dangling bonds*. A dangling bond is an unsatisfied valence on an atom, creating an immobilized, highly reactive free radical. Given the predicted density of dangling bonds ($\approx 5 \cdot 10^{-4}$ per nm^2) and the transistor's surface area ($\approx$ 80 nm^2), a 5 nm $\times$ 5 nm lane would have a 4.2% chance of failing, leading to an overall architecture breakdown. As a response, the industry moved on to create advanced 2.5/3D transistor designs which are more expensive to fabricate [42]. In 2.5/3D arrangements, silicon dies are stacked or placed side-by-side on top of an interposer, incorporating through-silicon vias through which electrically interconnects are established. Although being incredibly useful for the design of integrated circuits with nano-scale features fabrication, their high fabrication cost caused the cost of a transistor in 2014 to increase for the first time in history (**Figure 2.2**).

2.1.2 Processing speed

Another way around this issue is to use a higher clock rate to achieve a higher processing speed, as it dictates the number of operations the computer can perform per unit of time. However, as we track the advancement in CPU clock frequency, it seems like this process was almost halted in 2002 with Intel's Xeon 3.2 GHz chip (**Figure 2.3**). The reason for not using a higher frequency is anchored in our limited capacity to dissipate heat. The generated heat p is a function of the capacitive load (number of transistors) c, voltage v and clock frequency f according to: $p = c \cdot v^2 \cdot f$. Increasing clock frequency would, therefore, have a dramatic impact on the magnitude of the generated heat. Unfortunately, given that heat must be dissipated from a chip, which is ≈ 1.5 cm on the side, we have reached the limit of what can be cooled by air [117]. As a result, advanced cooling mechanisms were developed, including water cooling systems and even liquid nitrogen cooling kits. These can help overclock processing cores but they are bulky, expensive, require maintenance, have a relatively high failure rate, and have their own cooling restrictions. Today, thermal design is at the core of many architectural innovations. For example, when Apple introduced their MacBook Pro laptop design, they highlighted their "advanced thermal architecture" which enables better computing performance: "With large impellers, improved fan blades and additional heat-dispersing fins, the MacBook Pro delivers up to 12 watts more sustained power." Mechanical engineering and heat dispensing techniques currently stand at the front line of computing performance.

2.1.3 Distributed computing

Computing performance continues to improve, mainly by moving to multi-core designs. Due to the limits of parallelism dictated by *Amdahl's law*, computing performance kept improving but at a slower rate. Amdahl's law dictates some limitations to the number of available cores per chip. Suppose 20% of the algorithm cannot be distributed for multi-core processing. In that case, the maximum gain in performance is limited to $5x$, no matter how many cores are available. Along with the demise of Moore's law, Amdahl's law contributed to the fact that since

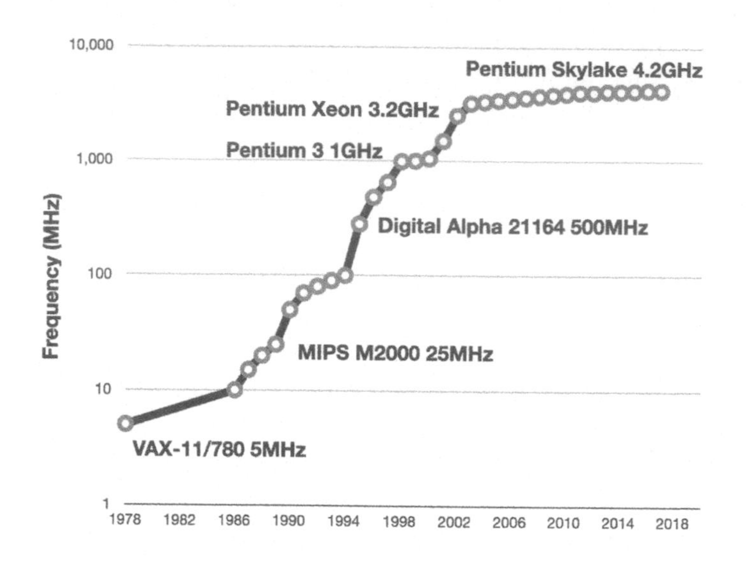

Figure 2.3 CPU's clock rate has been improved at an annual rate of 40% per year between 1986 and 2003. Since then, clock rate improvement has stagnated and is growing at an annual rate of 2% per year. Data from [117].

2015, growth in computing power is estimated at an annual rate of 3.5% (**Figure 2.1**). One particularly appealing technology for distributed computing is using GPUs for General Purpose (GP) computing (termed GPGPU). GPU computing has become an integral part of mainstream computing [222], providing thousands of computing cores working simultaneously on a given task. The importance of GPU computing for high-performance computing is evident in various fields ranging from graphics [218], computer vision [114], and machine learning [1] to brain simulations [158]. This is also evident in the rapid increase in GPU performance. For example, NVIDIA's V100 GPU has thousands of computing cores, tens of billions of transistors and was used to break the 100 tera FLOPS (10^{12} FLOPS) barrier of deep learning performance. However, this high-performance level comes with an energy toll of 250 to 300 watts. Such GPUs usually come with their own air or water cooling systems.

2.2 EMERGING COMPUTING PARADIGMS

The path toward improvement in computing performance led by industry, discussed above, was destined to hit a fundamental wall. *Carver Mead*, one of the founders of neuromorphic engineering, testified regarding this strategy: "Industrial practice was on a problematic path: in the race to release new product generations, it was faster to scale old designs to smaller feature sizes than to innovate at the architecture level" [201].

Facing this immense challenge in computing advancements, different computing paradigms have emerged in recent years, trying to radically change the computing paradigm. Some of these new paradigms show great promise for the future of computing. A far-from-exhausting list of emerging computing strategies includes

- Quantum computing
- Molecular computing
- DNA computing
- Programmable microfluidics

Another emerging computing paradigm is brain-inspired or **Neuromorphic** systems for which the primary rationale from the perspective of the computer architect will be described in the next section.

2.2.1 Quantum computing

Universal quantum computers leverage the quantum mechanical phenomena of superposition and entanglement to provide efficient means for computing. Quantum computers can solve quantum-related problems very efficiently. For example, in 2019, a group of scientists from Google and various academic institutions demonstrated the supremacy of a quantum computer for a particular quantum-mechanics related problem (random quantum circuit sampling) [22]. In this article, the authors argued that their quantum computer could solve this problem in 200 seconds while estimating that a supercomputer would take 10,000 years to accomplish the same task. As a side note, one month later, researchers in IBM argued that this was an overestimation of quantum supremacy, as their supercomputer in Oak Ridge National Laboratories would be able to complete that task in 2.5 days [225].

Nevertheless, quantum computers hold great promise for computing performance. This is true not only in terms of computing performance but also in terms of energy efficiency. *Salvatore Mandra* and colleagues implemented their simulator of Random Quantum Circuits (RQC) on both quantum and state-of-the-art supercomputers. They found that while the energy that a supercomputer requires to execute this simulation is 20 to $100 \cdot 10^6$ watts, a quantum computer requires $4.2 \cdot 10^{-4}$ watts, thus demonstrating orders of magnitude improvement in energy consumption [291].

How universal can quantum computers become? The authors of [22] left the readers with the statement: "We are only one creative algorithm away from valuable near-term applications." However, although promising in terms of computational capability and its expected improvement rate, quantum computers are still out of reach for most applications.

2.2.2 Molecular computing

In molecular electronics (for the sake of the discussion, DNA - DeoxyriboNucleic Acid- computing is discussed separately), logic and memory devices are fabricated from an ensemble of molecules, offering the potential to reduce device size, fabrication costs, and heat generation by several orders of magnitude. Various designs were proposed to realize molecular electronics. One approach was to create devices comprised of self-assembled programmable $1\mu m^2$ nano-cells, made out of metallic particles, with embedded molecular switches (nitro-containing oligo(phenylene ethynylene)s, sandwiched between metallic contacts). These nano-cells can be modulated in real-time, enabling programmable behavior [282]. Programmable carbon nanotube-based memory cells were also proposed [251].

While molecular electronics found their way into displays and single-molecule light-emitting diodes, they are currently far from being considered a potential replacement for silicon-based transistors. In their article: "The booms and busts of molecular electronics," *Kevin Kelly* and *Cyrus Mody* describe some of the promises and failures in molecular electronics [154]. They summarized it as a pursuit of an impossible dream.

2.2.3 DNA computing

A subclass of molecular computing is DNA computing. Driven by the rise of synthetic biology, DNA-based circuits were used to implement:

programmable dynamical systems [264]; nano-robots capable of targeting payloads to cells according to some logic-driven activation [278]; logical gates [254]; and memory [16]. Moreover, DNA computing was reported to be able to address NP-complete problems (where computing time grows exponentially with problem size) [178], and to perform digital computing [298]. In a fascinating work by *Erik Winfree* and colleagues, programmable DNA assemblies were shown to solve various algorithms, ranging from sorting and recognizing palindromes to electing a leader and simulating cellular automaton, thus, suggesting that self-assembled DNA entities could comprise a reliable computing device with a high-level of abstraction [307].

According to the National Human Genome Research Institute, the cost efficiency for DNA sequencing has improved faster than Moore's law, i.e., in 2020, full human scale DNA sequencing cost less than $1,000: a dramatic improvement considering the estimated cost of $100,000,000 in 2001. As the capacity to efficiently and affordably read and write DNA increases, the potential of using DNA for general computing seems to be a realistic endeavor. However, DNA-computing is still far from being considered comparable with digital computers in terms of programmability, usability, and reliability.

2.2.4 Programmable microfluidics

Microfluidics is a rapidly growing field with applications ranging from soft robotics [238] to quantum physics [263]. Principally, microfluidics is the science of the precise manipulation of fluids at a micro- to pico-liter scales [283]. A particularly interesting emerging technology that has evolved from continuous microfluidics is programmable microfluidics, often implemented with Programmable Valve Array (FPVAs) [93]. An FPVA is a dense grid of switchable blocks, with which fluid can be manipulated in highly configurable and programmable patterns. An FPVA provides a standard microfluidic architecture that can be configured to support nearly any relevant (mostly biological) application. Microfluidics large-scale integrated devices may have many thousands of such switchable blocks (or integrated micromechanical valves) and control components [277]. Interestingly, it has been shown that micro/nanofluidic systems follow Moore's law, as valve densities have increased exponentially with time [128], reaching 1 million valves per cm^2 [20]. While holding great promise in the realm of "Lab on a Chip" technology, microfluidics cannot currently compete with silicon-based devices' density and performance.

2.3 BRAIN-INSPIRED HARDWARE

The general idea of using the brain as *"a guiding principle"* to computer design, implementing distributed computing and in-memory computation, has its roots in the works of *Alan Turing* [286] and *Frank Rosenblatt* from the 1950s [246]. However, neuromorphic engineering as a field really kicked off when *Carver Mead* started to emulate neurons with transistors and *Bernard Widrow* developed the memistor.

When neuromorphic integrated circuits are discussed, all roads lead to Carver Mead. Back in 1981, *Carver Mead, Richard Feynman* and *John Hopfield* joined forces to offer a new graduate course: "Physics of Computation" at the California Institute of Technology. Carver Mead described the course as "exhausting, exhilarating and enlightening; the students were amazed, confused, and overwhelmed — and many of them have gone on to do incredible things" [200]. A few years later, the course concluded with Richard Feynman leading quantum computing [92], John Hopfield developing a new important type of neural networks [130] and Carver Mead designing neuromorphic systems. In his seminal work: "Neuromorphic electronic systems" published in 1990 [201], Carver Mead argues that "the brain is a factor of 1 billion more effective than our present digital technology and a factor of 10 million more efficient than the best digital technology that we can imagine." Seeing the approaching demise of Moore's law, Carver's graduate students innovated in many neuromorphic engineering fronts, finally starting their own groups and founding the field. Among them was *Misha Mahowald* who designed a neuromorphic chip which generated a disparity map (distance to objects) from stereo neuromorphic retinas. The *Misha Mahowald Prize* is annually awarded to "outstanding research in neuromorphic engineering," in memory of Misha, "one of the most influential pioneers of the field of neuromorphic engineering" (https://mahowaldprize.org).

In the following three sections we will discuss the why, how, and what of neuromorphic engineering.

2.3.1 The why

Modern digital computers are operated at a frequency scale of 10^9 Hz and are characterized by a power density of $\approx$ 50 watts per cm^2. In contrast, the brain operates at the $\approx 10^1$Hz regime with a power density of 0.01 watts per cm^2 [205] (**Figure 2.4**). Despite the remarkable difference in power density, the brain is postulated to work at the peta to exaflops regime ($10^{15} - 10^{18}$ FLOPS), a comparable performance to a state-of-the-art supercomputer, such as IBM's BlueGene which operates at the petaFLOPS regime (10^{15} FLOPS/second) (see detailed discussion in Section 4.2). Another way to look at it is this: if we scale the BlueGene to perform at the exaFLOPS regime, it would consume 20 megawatts of power. This is a million times the energy the human brain consumes.

Let's consider a typical robotic car participating in DARPA's grand navigation challenge for autonomous vehicles. This car would use a Global Positioning System (GPS)-defined path carrying over a kilowatt of sensing and computing power. Another (autonomous) machine is the honeybee. The honeybee would navigate its way back home with less than a million neurons powering its cognition, consuming merely a 10^{-3} watt of power. This is a millionth of the energy consumed by the car [179]. Moreover, the honeybee can generalize its computing capacity for other challenges, such as social interactions, and even thinking abstractedly about numerical concepts [131]. The honeybee has a *general intelligence*. How do 10^6 spiking neurons achieve this?

2.3.2 The how

The two main paradigm shifts in neuromorphic computing are:

- Probabilistic representation
- In-memory computing

Each of them will be briefly addressed next.

2.3.2.1 Probabilistic representation

In his article "A neuromorph's prospectus," *Kwabena Boahen* outlines the strategy the brain takes to gain its efficiency. The underlying principle for information communication, in accordance with *Claude Shannon's* information theory, is that the number of encoded bits of information b decreases logarithmically with the signal E to noise N ratio: *SNR* (assuming a communication channel with additive, white Gaussian noise):

$$b = \frac{1}{2} log_2(SNR) \tag{2.1}$$

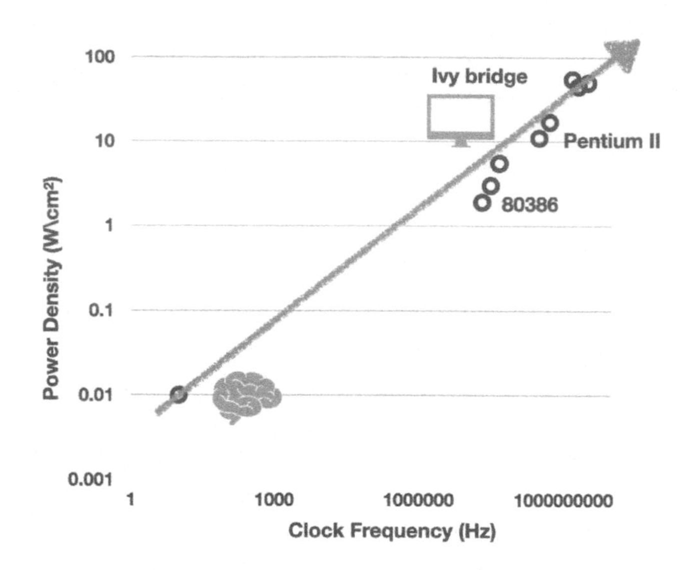

Figure 2.4 Traditional computing architectures took the faster way toward improvement by scaling up existing designs. This has resulted in increased power densities and clock frequencies. Optimized over many years of evolution, the brain operates at a much slower rate and power density while offering comparable computing performance. Data from [205].

Note that in the binary digital computer, b is capped to 1. Communication energy efficiency can be defined as:

$$E_{eff} = \frac{b}{E+N} \tag{2.2}$$

For example, if the SNR drops from 15 to 3, the number of encoded bits will drop from 2 to 1, and energy efficiency will double from $\frac{1}{8}$ to $\frac{1}{4}$. Hence, *SNR* can be traded for energy efficiency. This however will come with a price: increased error probability which can be defined by

$$ERR_p = e^{-0.25 \cdot SNR} \tag{2.3}$$

Equation 2.3 refers to the probability that a (thermal) noise would induce an error in a binary-coded data with a threshold of $\frac{1}{2}$. Factor 0.25 is used since an error is induced when the $SNR < 4$, or the noise's energy exceeds a quarter of the signal's energy.

A digital computer would invest an immense amount of energy generating a signal, achieving an *SNR* of 220, thus gaining an error probability of 10^{-24}. This is an incredibly small number which allows a modern 3.2 GHz Intel i9 processor with its 3.5 billion transistors to make a mistake once every $\approx$ 24 hours. However, for this high level of performance, the processor would have to pay with low energy efficiency of $\approx$ 0.0045 bit/N. In contrast, the brain invests a small amount of energy to generate a signal, achieving an *SNR* of only 1.72, thus gaining an error probability of 0.65. Indeed, when a spike is driven to a synapse to trigger a response in the postsynaptic cell, there is a 65% chance it will fail (**Figure 2.5**). The brain compensates for it with extensive connectivity. Under the same conditions, however, the brain will benefit an energy efficiency of 0.265, $\approx$ 60 times the efficiency of the digital computer (**Figure 2.6**). The bottom line is that the brain is efficient because it allows itself to make mistakes.

In his seminal work, *"Intelligent Machinery,"* *Alan Turing* disputes *Kurt Godel's theory*, according to which a machine "must not make mistakes" [286]. Turing argued that not making mistakes is not "a requirement for intelligence." On the contrary, Turing gives examples which show that making mistakes is a testimony of intelligence, as it provides the flexibility needed for robust and modular behavior.

2.3.2.2 In-memory computing

Another key paradigm shift in neuromorphic hardware design is the concept of in-memory computing. When general-purpose digital computers are utilized to simulate spiking neural networks, they are limited by the memory-CPU bottleneck (or memory wall), dictated by the von Neumann architecture. The von Neumann architecture separates the processor from the external memory, creating a data bus on which packets of information move at high-speed [205] (**Figure 2.7**). This bottleneck is partially relaxed in advanced architectures by using hierarchies of memories [117].

In the brain, memory, computation, and communication are tightly integrated and distributed over a communication fabric [205]. Guided by this architectural principle, neuromorphic designs typically combine memory, processing units, and communication in distributed modules which asynchronously communicate with each other via a shared bus (**Figure 2.8**). In this spike-driven framework, each module is

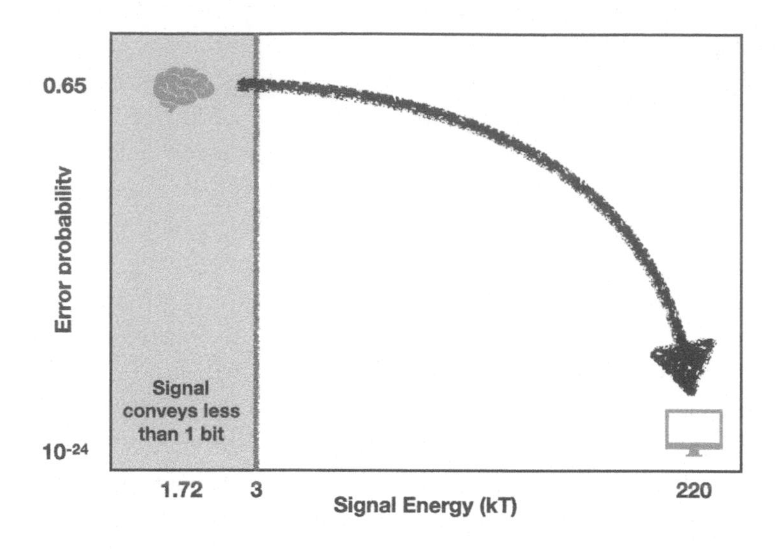

Figure 2.5 While a digital computer would invest 220 kT for each signal it generates, the brain will invest only 1.72 kT. This resulte in an error probability of 10^{-24} for the computer and 0.65 for the brain. When signal energy is less than 3 kT, the signal carries less than 1 bit of information. kT stands for thermal noise. Data from [42].

independent, not synchronized by a single clock, thus presenting an entirely new computing paradigm.

2.3.3 The what

To explore the range of applications for which neuromorphic hardware can be useful, it would be convenient to measure performance with representational accuracy. While energy consumption for the generation of an analog signal scales quadratically with precision (energy consumed to generate voltage scales quadratically with its amplitude), a digital signal scales logarithmically (adding a bit would multiply its representational capacity by 2). The crossover point in which digital representation becomes more efficient than the analog alternative is migrating to the left with improved fabrication. A neuromorphic computer which utilizes analog circuitry for computation and digital computing for communication is argued to achieve linear correlation between precision and energy consumption, thus achieving superior performance over five scales of precision [42] (**Figure 2.9**). While simple computational tasks are more efficiently executed with analog computers, exact applications would

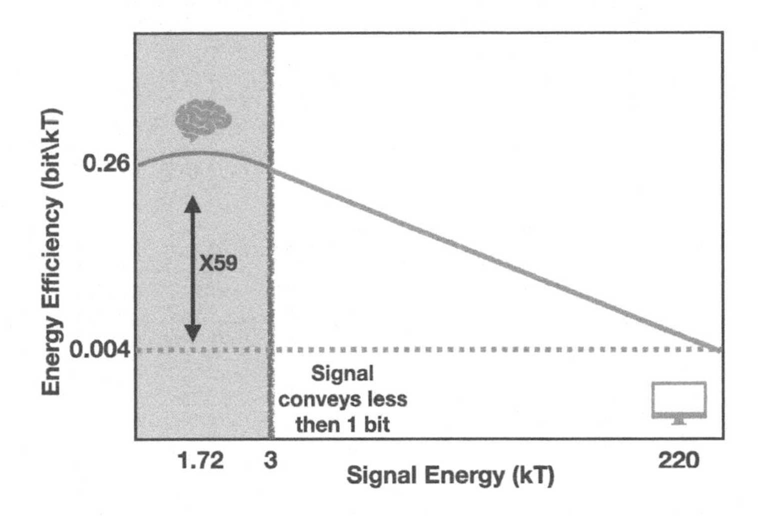

Figure 2.6 While a digital computer would invest 220 kT for each signal it generates, the brain will invest only 1.72 kT. This resulte in an energy efficiency of 0.26 for the brain and 0.004 for the computer. The brain is therefore 59 × more efficient than the computer. kT stands for thermal noise. Data from [42].

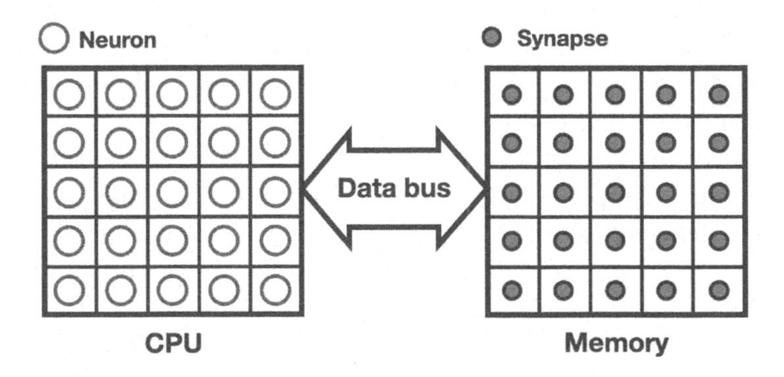

Figure 2.7 The von Neumann bottleneck. The von Neumann computer architecture dictates a separation between the processor and the external memory, resulting in the memory-CPU communication bottleneck.

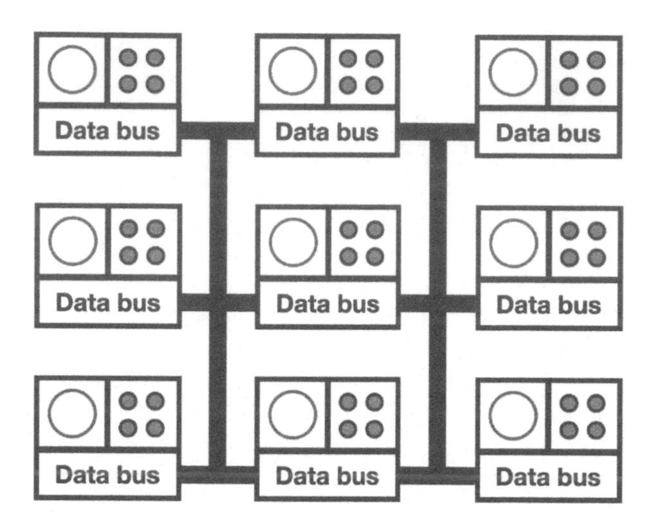

Figure 2.8 Conceptual neuromorphic architecture. A neuromorphic approach for computing is based on in-memory computing, where memory, computation, and communication are tightly integrated and distributed over a communication fabric.

benefit from the digital computer's superior representational characteristics. Between these two extremes, a vast array of applications, ranging from neuro-robotics to vision processing, can significantly benefit from neuromorphic hardware. While this analysis was derived for a hybrid analog/digital neuromorphic design, companies like IBM and Intel are approaching the neuromorphic realm with pure digital architectures. While pure digital designs are not as efficient as analog or hybrid designs, neuromorphic digital circuits still exhibit impressive power efficiency. For example, it was shown by *Chris Eliasmith* and colleagues that a robotic adaptive control, implemented on Intel's neuromorphic hardware (the Loihi chip), achieved superior performance in terms of accuracy with a power cost ≈ 5 to ≈ 40 times lower than its equivalent implementation on a CPU or a GPU, respectively [73].

To conclude, neuromorphic computing is not suitable for every task. It is overqualified to deal with low precision applications for which analog computing devices have the upper hand nor for exact applications for which digital systems prevail. Therefore, it is reasonable to believe that neuromorphic devices will complement existing digital frameworks, overtaking the challenges for which they offer clear advantages. That

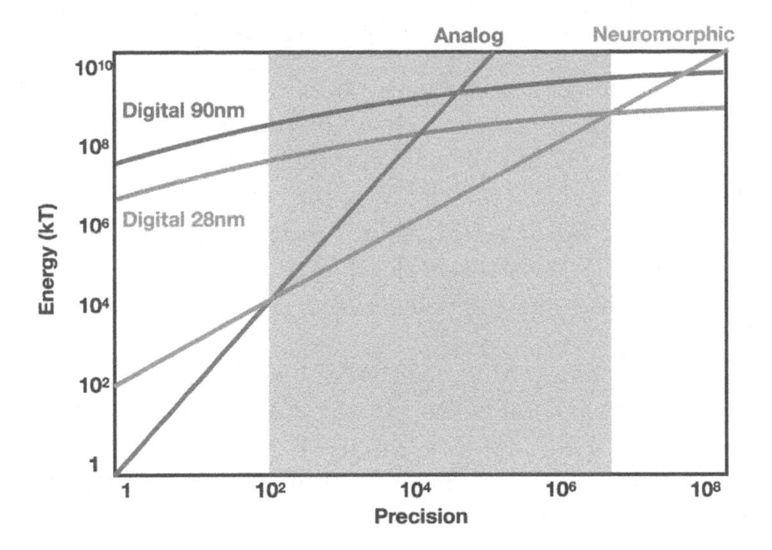

Figure 2.9 Signal energy and precision. While energy consumption for analog signals scales quadratically with precision, its scales logarithmically for digital signals. An neuromorphic hybrid analog-digital computer is argued to achieve linear correlation, achieving superior performance over 5 scales of precision. Data from [42].

being said, the benefits of neuromorphic hardware as accelerators over conventional neural networks accelerators are yet to be proven [80].

2.3.4 Neuromorphic frameworks

For the sake of discussion, I will divide the realm of neuromorphic frameworks into four main categories:

- Neuromorphic sensing
- Neuromorphic interfaces
- General purpose neuromorphic computers
- Memristors

Each of them will be briefly addressed next.

2.3.4.1 Neuromorphic sensing

Some of the first and greatest successes in neuromorphic computing architectures have been in vision and sound processing [137], including the silicon cochlea [183] and the silicon retina by *Misha Mahowald* and *Carver Mead* [187], and its biologically plausible version by *Kareem Zaghlou* and *Kwabena Boahen* [312]. More recently, neuromorphic Electromyography (EMG) sensors were developed. In general, neuromorphic sensing comprises the following categories:

- **Vision processing.** Most neuromorphic vision sensors communicate transients in luminance via the Address Event Representation (AER) (AER) protocol (discussed in Section 9.1). They are comprised of an array of silicon neurons; each generates spikes in response to a change in luminance in one particular location (or pixels). Spikes are time-multiplexed over an asynchronous data bus via an address encoder which designates each spike with a corresponding address (usually, the neuron's x-y pixel coordinate). These frame-less and event-driven neuromorphic Dynamic Vision Sensors (DVS) can resolve thousands of frames per second, have a fine temporal resolution, high dynamic range, and high signal to noise ratio. Moreover, since DVS perform sensor-level data compression, they optimize data transfer, storage, and processing, thus improving the power and size efficiency of the system [231].

- **Sound processing.** The human's cochlea turns vibrations in airpressure into spikes. Particularly, as a sound wave is introduced into the cochlea, it initiates a pressure wave that propagates along the Basilar Membrane (BM) length from base to apex. Different sections along the BM respond with spikes to different vibration frequencies, thus transforming sound to spikes. Following biology, silicon cochlea usually implement filters with a characteristic frequency - a preferred frequency stimulus. A detailed description of these circuits and variants is given in [179].

- **Muscle tone sensing.** Electromyography (EMG) signals correspond to muscle tension, thus reporting on movements. Neuromorphic EMG sensors were shown to classify in real-time muscle activity patterns such as hand gestures (opening and closing hands), reporting on a particular movement pattern with AER [78].

Neuromorphic sensors transform continuous reality into discrete events, thus being the bridge from the analog world to spike-tailored computing. Therefore, they pave the way to neuromorphic interfaces between man and machine, which will be the next topic of discussion.

2.3.4.2 Neuromorphic interfaces

Advances in both traditional machine learning methodologies and neuromorphic sensing are paving the way to new human-machine interfaces, where traditional touch-pads and joysticks are being replaced with intimate and multi-directional sensing capabilities [314]. Applications like artificial retinas, cochlear implants, and bionic prosthesis might gain tremendous benefits from advancements in neuromorphic sensing and their further development into man-machine interfaces, for example, a neuromorphic model of a reflex, implemented using EMG signal acquisition, was shown to improve the grasping capabilities of a prosthesis arm, with better adaptability to deformable, irregular, or heavy objects [219]. Moreover, to design advanced human-machine interfaces, hybrid neuromorphic acquisition of biological signals may be required. For example, a fusion of neuromorphic EMG sensing and neuromorphic vision was used to design a gesture recognition system with better-discriminating capabilities (20 to 40% decrease in inference time), and more efficient power performance [54].

2.3.4.3 General purpose neuromorphic computers

In the past two decades, numerous computing neuromorphic frameworks were developed, both by industry and academia. Some of the most notable circuits are the NeuroGrid, the SpiNNaker, the TrueNorth, and the Loihi:

- The **NeuroGrid** was designed in Stanford University by *Kwabena Boahen* and colleagues [33]. It is a hybrid analog/digital neuromorphic chip, in which computing is held in analog circuitry, and communication is managed digitally. While the NeuroGrid is programmable and exhibits efficient energy consumption, its design's main aim was to accelerate biological modeling. The board's utilization for general-purpose computing was limited due to its lack of inherited support for a computing framework with which spiking models can be employed for computation. More importantly, the analog part of the chip is harder to scale and non-trivial to configured with high-level programming. The NeuroGrid was morphed into different circuit architectures which addresses these issues, among them the BrainDrop [217].

- The **SpiNNaker**, was developed at the University of Manchester, by *Steve Furber* and colleagues [96]. It is fully digitized neuromorphic hardware, featuring a million-core computing engine whose goal is to simulate a billion neurons' behavior in real-time. The SpiNNaker is comprised of an array of interconnected ARM-based computing cores which are arranged in arrays of identical, independent, functional units to achieve distributed computing. These computational cores communicate with an interconnect fabric which has a bandwidth of $\approx 5 \cdot 10^9$ packets/s.

- The **TrueNorth** was developed by researchers in IBM, led by *Dharmendra Modha* [72]. It is scalable neuromorphic computer architecture in which 4,096 computational cores are interconnected. Each core operates in a parallel, distributed, and semi-synchronous fashion, receiving a 1 kHz clock for a discrete neuron update every 1 ms sec. Overall, the TrueNorth features 10^6 neurons and $256 \cdot 10^6$ synapses. It has a peak performance of $58 \cdot 10^9$ Synaptic Operations Per Second (GSOPS) and energy efficiency of 400 GSOPS/W.

- The **Loihi** chip was developed at Intel Labs [69]. Each Loihi chip is comprised of 128 neuron-cores; each core simulates 1,024 neurons and has 4,096 ports. Each chip also has x86 cores which are used for spike routing and monitoring. The chip integrates features such as hierarchical connectivity, dendritic compartments, synaptic delays, and programmable synaptic learning rules.

This is by no means an exhaustive list of all available neuromorphic hardware as new designs are proposed frequently and existing designs are morphs into more capable forms. However, it portrays the spectrum of available approaches, ranging from pure analog (sensing circuits) to hybrid and digital circuits. These designs, however, also share one primary concern: supporting high-level programming with no conventional operating system, non-deterministic communication, and almost no internal debug mechanisms. We therefore need a new programming paradigm in which functional spiking neural networks can be realized in underlying neuromorphic hardware with a high level of abstraction. We will explore some of the approaches developed for neuromorphic programming in Chapter 3.

2.3.4.4 Memristors

Another important approach for neuromorphic hardware is the development of the *memristors.* This field of research was initiated with *Bernard Widrow* developing the ADAptive LInear NEuron (ADALINE) in 1960 [304]. It is a 3-terminals device in which the conductance between two of its terminals is controlled by the time integral of the current in the third terminal. In 1970, *Leon Chua* conceptualized a generalized 2-terminal device, termed memristor, where the conductance between its terminals depends on the integral of the current driven through them, providing a synaptic-like behavior [58]. Memristors are portrayed as the "missing circuit element," referring to the existing three fundamental components: the resistor, the capacitor, and the inductor. Uniquely, the memristor's resistance depends on the past voltage waveform it has experienced. This form of memory-coupled computation can be used to design a neuromorphic learning network. Memristors were used to demonstrate spike-time-dependent plasticity and associative memory, thus providing hardware implementation of synapses. Memristors have the potential to transform neuromorphic engineering, making it genuinely scaleable. While memristors are not the focal point of this book, they will be succinctly discussed in Chapter 10. A rigor description of the memristor is available in [6].

2.4 GLOSSARY

In-memory computing: A computing architecture in which memory, computation, and communication are tightly integrated and distributed over a communication fabric.

Moore's law: Anticipating an exponential growth in computing power at an annual rate of 50% (doubling performance every two years).

von Neumann architecture: A prominent computer architecture in which the processor is separated from memory, creating a memory wall.

Transistor: Nano-scale switches for electrons in which electrons are driven from the transistor's drain to source and controlled by its gate's voltage.

2.5 FURTHER READING

- **Section 2.1**

- A comprehensive review of computer architectures, including analysis of computing advancement trends, are available in [117], a textbook by *John Hennessy* and *David A. Patterson.* A discussion of integrated circuits from the perspective of a neuromorphic engineer is given in [42].
- In [234], in *The Economist*, business opportunities which follow the demise of Moore's law are described. Emerging companies in the space of neuromorphic engineering (or the business of building brains) are described in [24].

- **Section 2.2**
  - Description of computational supremacy of quantum computers, jointly written by researchers in *Google AI Quantum*, *NASA Ames* and other academic laboratories was written by *Sunny Bains* in [22]. A survey of quantum computing technologies is given in [111]. As a side note, "Is the brain a quantum computer?" The short answer is "no" [177].
  - Early perspective on natural computing as a bridge between computer science and biology is given in [150] and an even earlier motivation in [249]. A review of bio-computing tools and aims is given in [152].
  - Implementation of logical gates and timers using microfluidic technology is demonstrated in [279], [55], and [232].

- **Section 2.3**
  - In [205], a comparison of digital and neuromorphic systems, in terms of clock rate and power density is given by IBM Research.
  - A comparative review of analog, digital, and neuromorphic systems is described in length in [42].
  - Power requirements of supercomputers used for large scale simulation of neural networks is available in [191] and more generally in [159].
  - A comparative analysis by *Travis Dewolf* of adaptive control in CPU, GPU, and neuromorphic hardware is given in [73].
  - A short history lesson on the emergence of neuromorphic hardware by *Carver Mead* is given in [200].

– Three classic papers on the development of neuromorphic engineering as a field are: [286], by *Alan Turing* on using the brain as a guiding principle for intelligent machines, [58] by *Leon Chua* on the development of the memristors, and [201] by *Carver Mead* on neuromorphic electronics.
– A biographic movie on the life of *Misha Mahowald*, one of the pioneers in neuromorphic engineering, and the visionary behind the silicon retina is available through dailymotion.com.

CHAPTER 3

Introducing the perspective of the algorithm designer

Abstract

Algorithms power the digital world. As machine learning algorithms aim to transform numerous fields of research and practice, the cross-section between brain sciences and algorithm design is growing in importance. This cross-section has been proven to provide extraordinary cross-pollination to both ends, including the developments of Deep Neural Networks (DNNs), Convolutional Neural Networks (CNNs), and Recurrent Neural Networks (RNNs). This chapter will introduce the perspective of the algorithm designer on neuromorphic engineering which aims to develop applications ranging from adaptive robotic control to object recognition with increased accuracy and precision. In this chapter, we will discuss neuromorphic software development environments and the rationale of using Spiking Neural Networks (SNNs) for algorithm design.

3.1 FROM ARTIFICIAL TO SPIKING NEURAL NETWORKS

3.1.1 Network architecture

Since the pioneering work of *Warren McCulloch* and *Walter Pitts* in 1943 [197], the development of the perceptron model by Frank Rosenblatt in 1958 [246], and the development of the backpropagation algorithm for training multi-layer perceptron in 1982 by *Paul Werbos* [301],

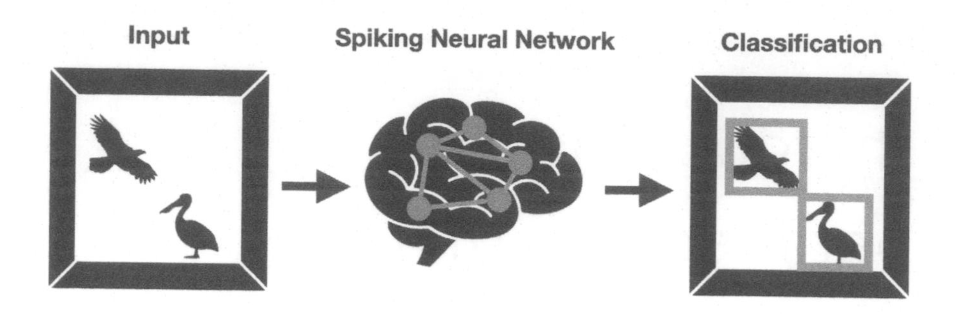

Figure 3.1 Neural networks can be trained to provide a classifier which categorizes an item into one of several categories.

Artificial Neural Networks (ANNs) have become an important cornerstone in machine learning, particularly through their utilization in different pattern identification and classification tasks (**Figure 3.1**). Although ANNs have roots in neuroscience, they are fundamentally different than Biological Neural Networks (BNNs), particularly in their learning mechanisms. ANNs can get deep with numerous interconnected layers and be trained over large data sets to provide a generalized model for predictive analytics. They are discussed in detail in countless books and they will not be the focal point for discussion here. A well-rounded guide to ANNs is available in [147].

While ANNs are typically trained prior to their evaluation with gradient-based optimization and regularization methods, BNNs are optimized in real-time via online-learning. In contrast to ANNs, which communicate differentiable values, SNNs propagate discrete spikes which, as was described earlier, can serve different data encoding strategies (Section 1.1). Spike-based computations are sparse in time, allowing better energy management and support of time-sensitive, noise-tolerated applications.

A model of a spiking neuron is comprised of a post-neuron driven by inputs pre-synaptic-neurons. Spiking activity is governed by a neuronal model which can features varying degree of biological plausibility, including the Hodgkin and Huxley model, the Izhikevich neuron model, and the Leaky Integrated-and-Fire (LIF) neuron model. These models will be discussed later in Chapter 5. From the perspective of the algorithm designer, the relevant abstraction for neural networks is neuronal spiking. Although the biophysical mechanisms for spike generations, cellular morphology, and synaptic types are fundamentally important from the perspective of the scientists, they are usually abstracted away in SNNs.

Like most ANN architectures, SNNs are comprised of spiking neurons interconnected via synapses through adjustable weighted lines, creating a network through which spikes are propagated. Within this network of interconnected layers of spiking neurons, spikes $x_i^{(1)}$ are modulated by synaptic weights, w_i to produce the resultant current: $\sum_{i=0}^{n_x} x_i w_i$, at a given time. In this proposed notation, the numeral 1 refers to the example number, as a network can be modulated with numerous inputs arriving at the networks in different time-windows; and n_x is the number of inputs (or features) connected to the neuron. This current affects the "voltage" of the post-neuron which follows certain neuronal dynamics. The neuron generates an outgoing spike whenever the neuron voltage crosses a certain threshold [248] (**Figure 3.2**). Network's weights optimization stands at the foundations of machine learning where during model training, the connection fabric of the network is optimized via weight updates to create a predictive model.

Spiking neurons can be organized in layers and connected to other neurons, constructing a deep SNN. Several deep SNNs architectures have been proposed. In a fully connected SNN, each neuron in one layer is connected to every neuron in the successive layer (**Figure 3.3**). In convolutional SNNs, convolution and sampling layers followed by a feed-forward classifier are concatenated to support image/vision-related tasks. In recurrent SNNs, a neuron in time t receives input along with feedback from its previous state in $t-1$, allowing it to find temporal patterns for applications, such as speech recognition and Natural Language Processing (NLP) (See Chapter 13 for a detailed discussion).

3.1.2 Neuromorphic applications

Neuromorphic applications are bound to semi-precise computation (See Section 2.3.3). They therefore excel in cognitive or pattern-driven applications, such as:

- Machine learning
- Control tasks (neuro-robotics)
- Cognitive models

Each of them will be briefly addressed next.

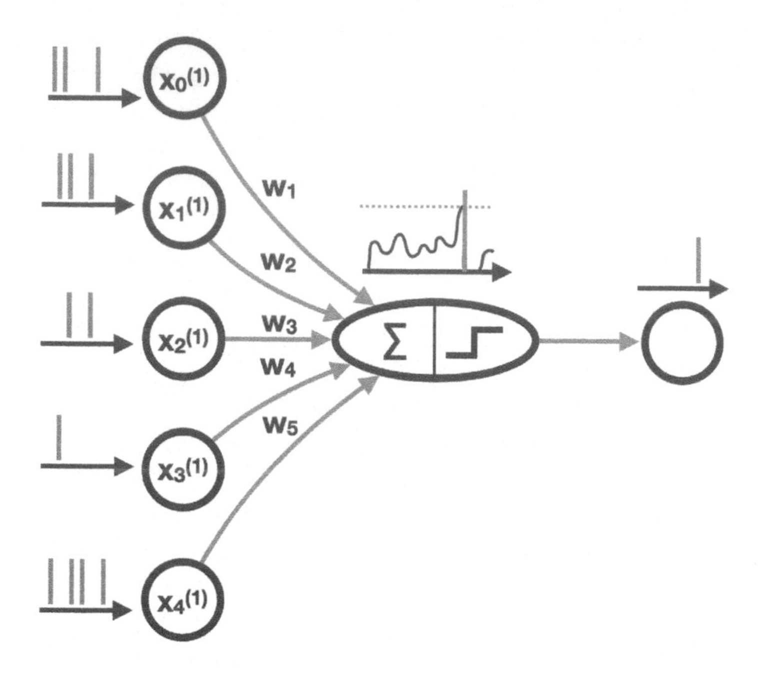

Figure 3.2 A spiking neuron produces spikes in a response driven by spikes arriving from other neurons. Incoming spikes are filtered, weighted, summed, and evaluated according to some neuronal dynamics which dictates the threshold for the initiation of output spikes.

3.1.2.1 Neuromorphic learning

SNNs can be trained with unsupervised, supervised, and reinforced learning methodologies. Learning with SNNs, from the perspective of the scientist, will be described in Section 7.4, and from the perspective of the algorithm designer, in Chapter 13. Briefly, neuromorphic unsupervised learning can be implemented with biologically plausible weight update rules, such as the Spike Timing Dependent Plasticity (STDP) [258] rule; supervised learning can be implemented similarly to ANNs, via weights optimization using neuromorphic back-propagation algorithms, such as SpikeProb [45]; and reinforcement learning can be implemented using real-time (or online) learning rules, such as the Prescribed Error Sensitivity (PES) rule [31]. Importantly, numerous studies concentrate on equivalenting ANNs and SNNs, enabling transfer learning - using weights learned in ANNs (with some modifications) as initial weights for SNNs [250]. Neuromorphic learning was shown to perform well in both vision

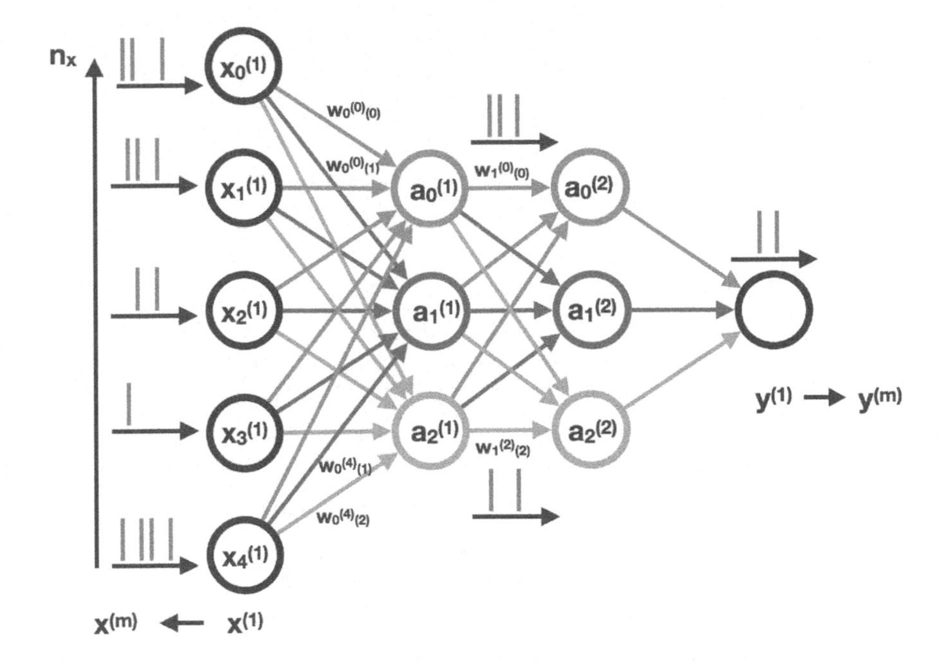

Figure 3.3 Spiking neurons can be organized in layers to construct deep neural networks, wherein each neuron in one layer is connected to each neuron in the successive layer. Each connection has an adjustable weight.

and sound processing. For example, NEF was used to design a neuromorphic algorithm which feature 94 % and 99.44 % on CIFAR-10, a database with 60,000 color images, classified into 10 classes (airplanes, birds, cats, etc.) and MNIST databases, a database with 70,000 samples of hand-written digits, commonly used as a benchmark for ANNs, respectably [250] [268]. Interestingly, a recurrent SNN which utilizes Legendre Memory Units (LMUs) was shown to outperform Long Short-Term Memory (LSTM) network, a state-of-the-art artificial RNN [296]. Particularly, LMU was shown to improve memory capacity and significantly reduce training and inference times for chaotic time-series prediction tasks. SNNs are in many ways becoming comparable in performance to ANNs.

3.1.2.2 Neuro-robotics

The control of robotic systems is currently dominated by Proportional, Integral, and derivative (PID)-based modeling. Such a model aims to accurately represent the controlled system, such that it would be able to capture the effect of external perturbations and its internal dynamics on the system's ability to move. Thus, it provides signals for movement control, given a desired location. However, in a human collaborative-assistive setting, the controller should consider kinematic changes in the system, such as object manipulation of an unknown dimension or at an unknown gripping point. For example, non-adaptive controllers are more likely to fumble or drop an object due to variations in their surface properties. Robot controllers are also less capable of adapting to new tasks or environments, often requiring extensive re-tuning of control parameters to accommodate new motions and surroundings. Neuromorphic control algorithms may acquire some of the advantages of biological motor control. Indeed, neuromorphic systems have been shown to outperform a PID-based implementation of the required non-linear adaptation, particularly in their ability to handle a high Degrees of Freedom (DOF) systems. Neuromorphic adaptive control utilizes online learning with SNNs to account for unexpected environment perturbations. These neuromorphic algorithms may closely emulate key features of neurophysiological analogs, such as cerebro-cerebellar inverse models [140] and vestibular and oculomotor circuits [169]. At the same time, these algorithms may selectively ignore or improve upon motor deficits of biological systems. Moreover, power comparison between different adaptive control implementations shows that conventional CPU and GPU require relatively more power for execution while having comparable latency performance [73].

3.1.2.3 Cognitive models

Ultimately, the brain's uniqueness lies in its ability to think abstractedly and creatively about the world. Our brain is capable of thinking in symbols, creating language, encoding something in memory and retrieving it by context (instead of by address), choosing one favorable action among available others, and representing time and space. It can do all of that without a centralized processing unit or register-based memory. To what extent can we implement this set of capabilities with SNNs?

In his book "How to build a brain," *Chris Eliasmith* attempts to address this question with the development of the SPA [83]. SPA uses circular convolutions to encode relationships between symbols (represented with vectors) and use it to solve problems like: the binding problem (Given a purple square and a pink circle, how can a SNN be used to perceive the square as purple and the circle as pink, and not the other way around?); interpreting homographs (symbols with several meanings, e.g., "she was sitting at the bar" and "I ate three bars of chocolate"); interpreting word types (nouns, verbs, adjectives, etc.); and representing space (binding properties to space-representing vectors). In 2014, *Daniel Rasmussen* showed that the SPA could use this set of principles to pass an intelligence test with human-like scores and error patterns [240]. As was briefly described in Section 1.2.2.2, SPA was used to design *Spaun*, a functional model of the brain, capable of responding to visual stimuli and executing a diverse set of tasks (e.g., digit recognition, tracing from memory, serial working memory, question answering, addition by counting, and symbolic pattern completion).

3.2 NEUROMORPHIC SOFTWARE DEVELOPMENT

Various frameworks for SNN programming were proposed over the past decade; some of them were widely adopted. One of the most widely accepted modeling frameworks in neuroscience is *NEURON* (by Yale University) [123]. NEURON aims at simulating highly detailed and biologically plausible models, and it was developed with its C-inspired Domain-Specific programming Language (DSL), called HOC. However, an interface between NEURON and Python was developed to allow for easier programming [125]. In Section 1.2.2, one of the flagship simulations of the Blue Brain project of a highly detailed model comprised of 31,000 cells of the cortical column [192] is described. This simulation was implemented in a NEURON-based environment which was specifically designed to be executed on supercomputers. To this aim, an extension of NEURON, named *CoreNEURON* was designed [161]. CoreNEURON allows an optimized distributed execution of a NEURON simulation with various computing architectures, including IBM's Blue-Gene/Q, CPUs, and GPUs. Another NEURON-tailored development environment is *NeuroConstruct* [103]. With neuroConstruct, 3D models can be defined with ideally little need for writing code and simulated with NEURON. NeuroConstruct have various extensions, including a framework for defining and simulating retinal circuitry [82].

Since simulating the exact dynamic of biologically plausible models requires an immense amount of computational resources, most implementations of SNNs abstract away most of the biological details (e.g., biophysical and morphological properties). For them, NEURON is not appropriate. *BRIAN* was developed as an alternative, easy to

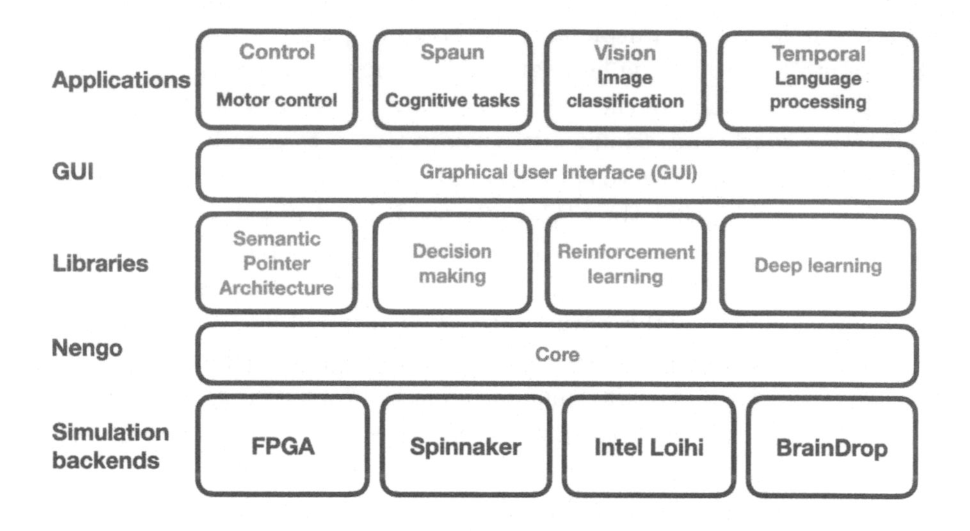

Figure 3.4 Nengo is a neural compiler that supports the compilation of high-level descriptions of SNNs for various neuromorphic hardware. In its back-end, Nengo provides execution code for BrainDrop, Loihi, SpiNNaker, and ABR's Field-Programmable Gate Array board [213]. In its front-end, Nengo provides a GUI and libraries, which extend the Nengo core to support different applications such as motor control, computer vision, and classification tasks.

use, Python-based library for SNN development, allowing the simulation of explicitly defined differential neuronal models [106] (See Chapter 11). Another powerful framework, which allows for high-level programming of SNN is the NEF, briefly mentioned in Section 1.2.2. *Nengo* is a Python-based software library which realizes NEF [32]. Nengo is essentially a neural compiler which supports the compilation of a high-level description of computation into its implementation on various neuromorphic hardware, including the SpiNNaker, TrueNorth, and the Loihi. Nengo is comprised of a Graphical User Interface (GUI)-driven front-end which provides a set of objects, including neuron ensembles, nodes, and connectors with which a NEF model can be defined. It also includes a back-end which provides an execution and hardware deployment framework (**Figure 3.4**). Nengo was extended with different libraries, providing algorithms for various tasks ranging from NLP to computer vision. NEF and Nengo will be discussed in Chapter 12.

Some software development environments are hardware-specific. One such framework, named *Corelet* was developed by IBM, to support their TrueNorth chip [11]. Corelet was designed as an object-oriented language to create corelets – abstract TrueNorth modules with exposed inputs and outputs and encapsulated neuronal details. IBM defined the Corelet Library as a repository of reusable corelets from which new corelets can be composed. The framework was deployed via the Corelet Laboratory which integrated Corelet with Compass, a simulator for large-scale models of TrueNorth cores, capable of simulating tens of millions of cores. IBM invested significant effort to create a software ecosystem which supports all aspects of the TrueNorth-based SNN programming cycle from design, development, and debugging to deployment.

3.3 GLOSSARY

Artificial Neural Network: Interconnected group of nodes, inspired by a mathematically abstracted model of a biological neural network.

Biological Neural Network: A neural network comprises biological neurons connected via electrical or chemical synapses.

Convolutional Neural Network: A deep neural network which incorporates convolutional and pooling layers. Mostly used for training over matrices (or frames) for visual processing.

Deep Neural Network: An artificial neural network wherein groups of neurons are organized in interconnected layers.

Recurrent Neural Network: A deep neural network that features feedback (or recursive) connections between neurons. Mostly used for training over temporal sequences.

3.4 FURTHER READING

- **Section 3.1**
  - A review of SNN architectures and training algorithms, along with performance measurements for different benchmarks is given in [273].
  - In-depth discussion of deep SNNs and convolutional SNNs is given in [133]. NEF-based implementation of deep SNN is

given in [250]. SNNs for visual processing, emphasizing transfer learning from ANNs with NengoDL, is given in [239].

  - SNNs for robotic control are reviewed in [37] and showcased for a robotic arm in [74], [73] and [64].
  - The SPA is described in length in [83], and its utilization for intelligence tests is described in [240].

- **Section 3.2**
  - The NEURON framework can be explored at neuron.yale.edu, CoreNEURON in [161], neuroConstruct in [103], BRIAN in [106], Nengo in [32] and IBM's Corelet programming environment in [11].

II

The Scientist's Perspective

CHAPTER 4

Biological description of neuronal dynamics

Abstract

Neuronal models are the backbone of neuromorphic engineering. They span a wide range of complexities, trying to maintain the delicate balance between bio-plausibility and model tractability. This chapter will discuss the fundamentals of the scientist's perspective on neuromorphic engineering, emphasizing the biological description of neuronal dynamics. The main aim of this chapter is to provide the necessary background to comprehensively understand the followed electrical and mathematical descriptions. The chapter will present a useful way to coarsely grasp some biological details which can later be utilized to design large-scale neuronal simulations and electrical implementations.

4.1 POTENTIALS AND SPIKES

As was described in Section 1.1, the neuron has the canonical description of being coarsely comprised of dendrites (signal input pathways), a soma (site of signals integration), and an axon (signal output pathway). Neurons typically communicate with (many) other neurons with spikes – temporary changes in voltage which propagate as impulses from the cell's soma through its axon to target neurons via synapses. In this chapter, the neuron's main features will be succinctly discussed. We will start by describing the maintenance of the neuron's resting potential, from which a spike or an action potential, can be initiated. We will briefly discuss the mechanism for the initiation of the action potential, its propagation through the axon, and its effect on the synapse and the postsynaptic cell.

4.1.1 The resting potential

A cell membrane separates the cell's interior (comprised of bio-molecules (proteins, DNA, etc.), specialized organelles, and structural filaments) from the environment. Particularly, the membrane separates populations of charged ions, where differences in the amount of charge on either side of the membrane create a potential difference or voltage. In a steady-state, where no net transport of ions through the membrane is apparent, the cell is at rest, and the membrane potential is termed a *resting potential.* What is the resting potential balancing? Two types of forces can drive ions across the membrane:

- A chemical force E_{ion} which drives molecules down their concentration gradient which, according to the *Nernst equation*, equals:

$$E_{ion} = C \cdot ln\frac{[ion]_{out}}{[ion]_{in}} \tag{4.1}$$

  where $[ion]_{in}$ and $[ion]_{out}$ are the ion concentration in and outside of the cell, respectively, and $C = 25.2$ mV at room temperature. This is an emerged entropic force, striving to have all ions homogeneously diffused.

- An electrical force, created from an unequal distribution of negative and positive charges across the membrane. In the cell, there are fixed ions which cannot move across the membrane and thus creating an electrical force, that is striving to keep positively charged ions inside the cell.

Concerning each ion, when the electrical force balances the concentration gradient force, there is no net transport of that ion. This is the ion's *equilibrium potential.*

To approximate the membrane's resting potential, we will define the current flow of an ion I_{ion} using Ohm's law:

$$I_{ion} = g_{ion} \cdot V \tag{4.2}$$

where g is the ion conductance through the membrane, defined as the inverse of the resistance $g = \frac{1}{R}$ (g has the units of Siemens S) and $V = V_m - E_{ion}$.

In neurons, the two main participating ions in the creation of the membrane potential are sodium (N_a^+) and potassium (K^+). Active channels $N_a^+K^+$ invest energy to drive N_a^+ ions out of the cell and K^+ into the cell, thus creating a *concentration gradient* (**Figure 4.1**).

In equilibrium, the flow of sodium into the cell $-I_{N_a}$ equals the flow of potassium out of the cell I_K, resulting in: $I_K = -I_{N_a}$. Referring to Equation 4.1, to the membrane potential V_m and to Equation 4.2, we derive: $-g_{N_a}(V_m - E_{N_a}) = g_K(V_m - E_K)$. Rearrangement yields:

$$V_m = \frac{(E_{N_a} g_{N_a}) + (E_K g_K)}{g_{N_a} + g_K} \tag{4.3}$$

In excitable cells (e.g., neurons and skeletal muscle cells), $E_K = -75$ mV, $E_{N_a} = 55$ mV, $g_K = 10 \cdot 10^{-6}$ and $g_{N_a} = 0.5 \cdot 10^{-6}$. The membrane potential of excitable cells, therefore, is approximately -70 mV (The inside of the cell is more negative than the outside). Note that the ion with the greatest conductance across the membrane at rest is potassium, and its equilibrium potential is, therefore, the major contributor to the resting membrane potential.

The above is by no means a complete description of the resting potential, as other ions and mechanisms contribute to the cell's membrane potential. For the purpose of this book, however, it is sufficient.

4.1.2 The action potential

Neurons generally propagate synaptic electrical inputs through their dendrites to the soma where they are integrated, causing the initiation of an *action potential* in the axon hillock. How is the action potential generated? In response to a stimulus, the cell membrane is depolarized, often increasing the membrane's conductance to sodium. The inside of the cell becomes less negative, resulting in increased membrane potential. Once a certain threshold voltage is reached (at ≈ -55 mV), an action potential is initiated. Particularly, at threshold value, voltage-dependent sodium channels open, allowing N_a^+ to diffuse into the cell. As a result, membrane voltage rises quickly to ≈ 30 mV. At this point, N_a^+ channels close and potassium channels begin to open. As K^+ diffuses out of the cell, re-polarization occurs and the cell membrane regains its polarization. This fast depolarization / re-polarization response to stimulus at the axon hillock constitutes the action potential or the spike (**Figure 4.2**).

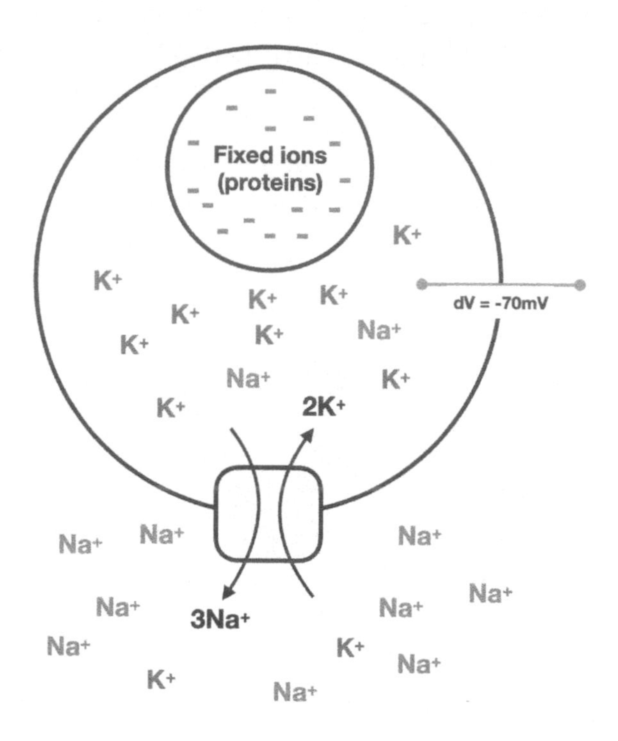

Figure 4.1 Equilibrium potential is maintained by the cell as it strives to keep the concentration gradient and electrical driving forces balanced at ≈ -70 mV.

4.1.3 Spike propagation

Once an action potential is generated at the axon hillock, it propagates down the axon. Briefly, when an axonal region produces an action potential and undergoes a depolarization, it serves as a stimulus for the axon's neighboring region. In this way, action potentials are regenerated along each small region of the axon membrane (**Figure 4.3**). Some axons are myelinated, i.e., they are coated with myelin created by supportive cells (myelin occludes N_a^+ channels). In myelinated axons, action potentials are produced at the *nodes of Ranvier* which are separated by highly resistive axonal regions. When one node depolarizes, it induces depolarization at the next node, causing action potentials to jump between nodes. As a result, in myelinated axons, action potentials move faster and farther.

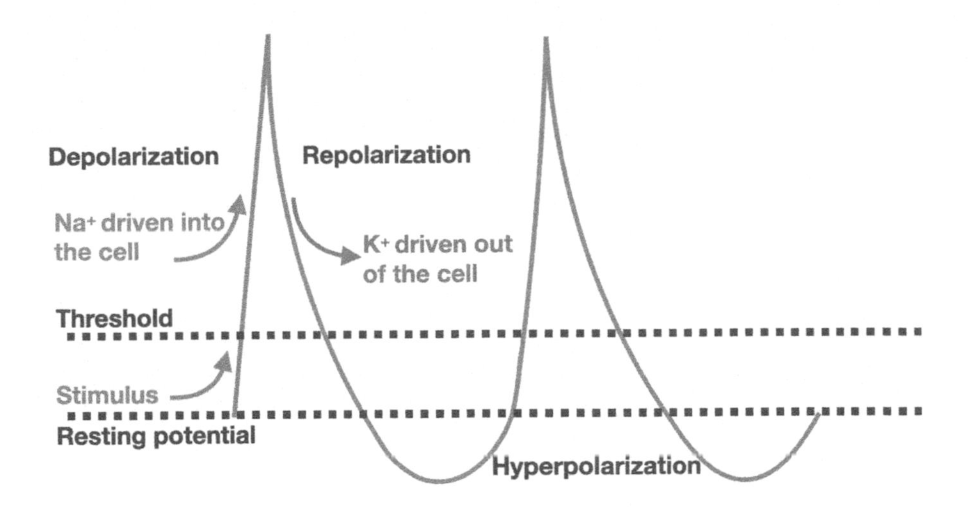

Figure 4.2 Action potential. In response to stimuli, the cell membrane is depolarized. Once a threshold voltage is reached, an action potential is initiated through the opening of N_a^+ channels and the diffusion of N_a^+ into the cell. As a result, the membrane voltage continues depolarizing until N_a^+ channels close and K^+ channels open. K^+ then diffuses out of the cell, causing re-polarization, followed by a voltage undershoot (hyper-polarization).

4.1.4 Synapses

Neurons are connected via synapses. Action potentials arriving at the end of an axon trigger the uptake of calcium (C_a^+) which causes *synaptic vesicles* to fuse with the axon terminals. Synaptic vesicles are small structures enclosed by a membrane and store neurotransmitters. Once a vesicle is fused with the axon terminal (buttons), its encapsulated neurotransmitters are released at the synapse. These transmitters diffuse across the synaptic gap and bind to receptors anchored on the postsynaptic membrane. Neurotransmitters induce ion flux across the membrane and can be either excitatory (by promoting membrane depolarization) or inhibitory (by promoting membrane hyperpolarization)(**Figure 4.4**). Being an inhibitory or excitatory synapse depends on the neurotransmitters encapsulated within the synapse's vesicles. For example, while the neurotransmitter glutamate would have an excitatory effect, Gamma AminoButyric Acid (GABA) would have an inhibitory effect. Modeling vesicle fusion dynamic and vesicle availability at the synapse is an

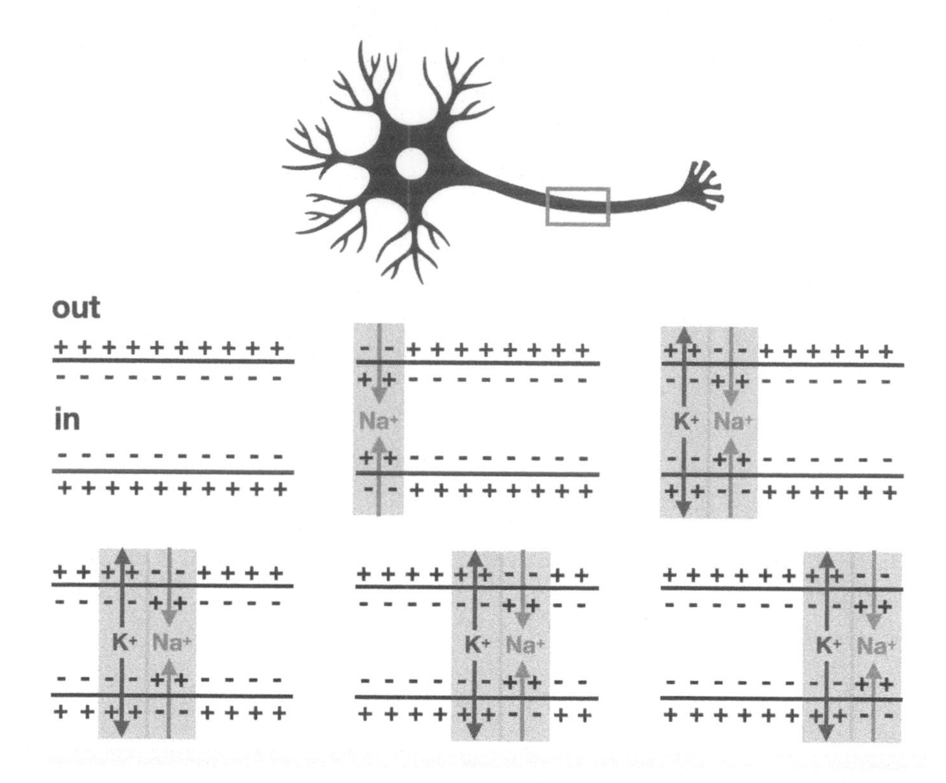

Figure 4.3 Propagation of action potentials. Once an action potential is generated, it propagates down the axon. When a region produces an action potential and undergoes a depolarization via an influx of N_a^+ ions into the cell, it serves as a stimulus for the next region of the axon. In this way, action potentials are regenerated along each small region of the axon membrane.

important modeling endeavor [28]. Biological synapses can therefore be either excitatory (positive "weight") or inhibitory (negative "weight"); they cannot change their encoding sign (Dale's law [67]). This poses a challenge for formulating supervised learning with biological synapses, as described later in Section 7.4.1.

4.2 POWER AND PERFORMANCE ESTIMATES OF THE BRAIN

There are many ways in which power consumption of the brain's elementary operations can be estimated. One way, which was proposed by *Kwabena Boahen* in [42], is the following: an ion channel opens and closes

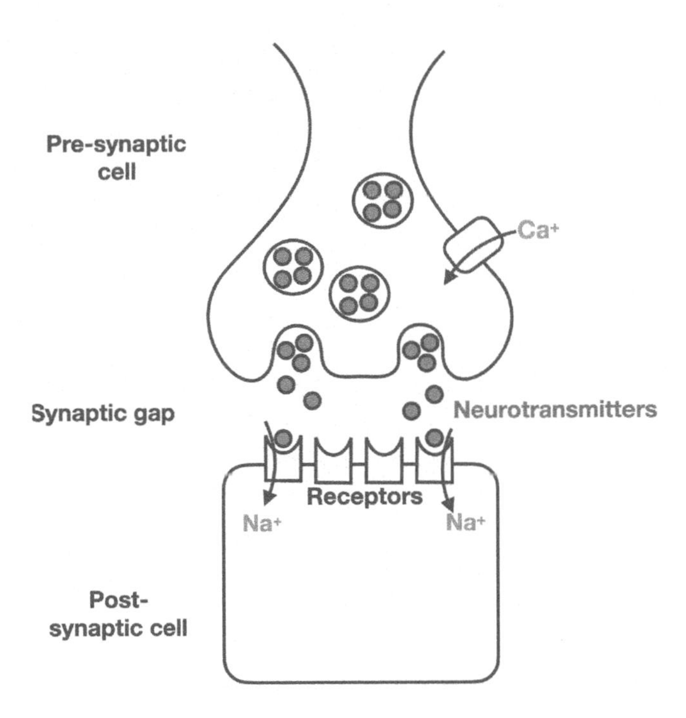

Figure 4.4 Neurons are connected via synapses. Action potentials arriving at the end of an axon trigger the uptake of C_a^+ which causes synaptic vesicles to fuse with the axon terminals, releasing their encapsulated neurotransmitters. These transmitters diffuse across the synaptic gap and bind to receptors anchored on the postsynaptic membrane, inducing ion flux, which causes depolarization/hyperpolarization.

continually, passing a current of 0 and $5 \cdot 10^{-12}$ A respectively. A spike reaching a synapse has a 20 % chance to open a channel. These channels close exponentially fast with a time constant of ≈ 10 ms. Thus, the ion channel conducts an average of $0.2 \cdot 5 \cdot 10^{-12} = 10^{-12}$ A across 0.1 V (approximately the resting potential). Each opening of an ion channel therefore consumes $10^{-12} \cdot 10^{-1} \cdot 10 \cdot 10^{-3} = 10^{-15} = 1$ fJ (Femto Joule). When communicating a spike, 20 ion channels are opened per synapse, resulting in power consumption of 20 fJ for synapse activation, or $20 \cdot 10^{-15}$ J per synapse. Therefore, to convey spikes from each of the brain's 10^{15} synapses, 20 J are consumed. Since spikes arrive approximately at a frequency of 1 Hz (1 spike/second/synapse) and 1 watt is defined as 1 Joule per second, a total of 20 W can be estimated as the brain's power consumption.

Another method proposed in [146] is based on metabolic rates. Considering a metabolic rate of 1,300 kilocalories (resting mode) per day, we estimate consumption of 15.04 gram calories per second. This can be converted to 62.93 Joules per second, which is about 63 watts. As the brain demands 20 percent of the resting metabolic rate, we reached $0.2 \cdot 63 \approx 12$ W.

To estimate the number of operations being evaluated by the brain in FLOPS, let us consider a rough estimate of 10^{11} as the number of neurons in the human adult's brain. The following chapter will cover some basic neuronal modeling methods; one of them is the Hodgkin-Huxley (HH) model (Section 5.3) which is considered a biologically plausible model. Assume that each neuron has 10^4 compartments, each executing the basic HH equations with 1,200 FLOPS [142], thus resulting in a total of $1.2 \cdot 10^{18}$ FLOPS. Each compartment would have 4 dynamical variables and 10 parameters, each described by one byte. We can, therefore, further estimate a memory capacity of $1.12 \cdot 10^{18}$ bits. Other estimates are summarized in [46].

4.3 GLOSSARY

Action potential: A brief electrical activity (a spike), created by a depolarizing/repolarizing currents of sodium and potassium ions across the membrane.

Resting potential: Relatively static membrane potential of quiescent cells in which no net transport of ions across the membrane is apparent.

Spike propagation: Propagation of action potential along an axon, driven by sodium channels and prevented by activation of potassium channels.

Synapse: A site of impulse (action potential) transmission between neurons, via neurotransmitter release.

4.4 FURTHER READING

- **Section 4.1**
  - In this chapter, a brief description of neuronal dynamics from a biological perspective was presented, enough to provide

the necessary background to understand the following chapters. The ideas described in this chapter are also given in khanacademy.org.

– A more in-depth discussion of neuronal dynamics is advised to the interested reader. One excellent resource is the book: "Neuroscience: Exploring the Brain" by *Mark Bear* and colleagues [30].

- **Section 4.2**

  – A deeper, region-oriented energy consumption of the brain is given in [280]. An analysis of energy consumption during neural computation is given in [255].

CHAPTER 5

Models of point neuronal dynamic

Abstract

Ada Augusta, countess of Lovelace, the daughter of poet Lord Byron, was a visionary English mathematician, considered by many as "the first computer programmer" for writing an algorithm for a mechanical computer ("analytical engine") designed by *Charles Babbage* in the mid-1800s. Ada was quoted back in 1815, saying:

> I have my hopes, and very distinct ones too, of one day, getting cerebral phenomena such that I can put them into mathematical equations. In short, a law or laws for the mutual actions of the molecules of brain... I hope to bequeath to the generations a calculus of the nervous system.

About 100 years later, *Louis Lapicque* provided one of the most widely utilized differential models for neural activity: the passive membrane model, later extended to the Leaky Integrated-and-Fire (LIF) model. Another important development is the Izhikevich neuron model, proposed by *Eugene Izhikevich* in 2003, offering a quadratic LIF with a recovery variable. A bio-plausible model was suggested by *Alan Hodgkin* and *Andrew Huxley* in 1952: The HH model, for which the authors were awarded the Nobel Prize in Physiology in 1963. The HH model provides a detailed differential description of neuronal dynamics and it is the principal mathematical description used for biologically plausible neuronal simulations. This chapter will discuss some of the mathematical formulations of these models to the level appropriate for this book's scope.

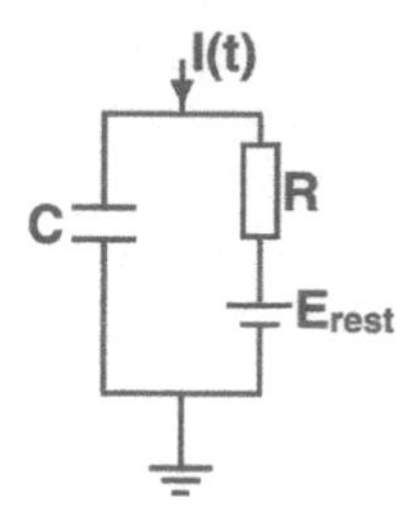

Figure 5.1 The passive membrane circuit model. The cell membrane acts as a leaky capacitor which models the membrane capability to separate ions. A resistor is used to model the ion channels.

Due to the importance of the LIF model, it will be more profoundly discussed.

In this chapter, I assume some familiarity with elementary electrical components and familiarity with differential calculus basics. The reader is encouraged to consult the Further Reading section.

5.1 THE LEAKY INTEGRATE AND FIRE MODEL

The LIF model is the simplest, most used electrical-mathematical neuronal model. We will start by defining the electrical model, explaining the neurons' dynamics, and then analyzing the model mathematically.

In LIF, the cell membrane acts as a leaky capacitor which models the membrane capability to separate ions (the passive membrane model) [2, 165]. A capacitor's voltage, in the absence of an injected current, exponentially decays (or "leaks") to a resting potential through a resistor which models the membrane permeability to ions (**Figure 5.1**).

Consider a neuron modeled with capacitor C, membrane resistance R and a small battery u_{rest} which constitutes the membrane's resting potential. Assuming the current is not "stored" or "lost," we can apply the following conservation:

$$I(t) = I_R(t) + I_C(t) \tag{5.1}$$

where $I(t)$ is the total current driven through the membrane in time t, $I_R(t)$ is the current driven through the resistor R, and $I_C(t)$ is the current driven through the capacitor.

A capacitor separates charges across its terminals in a correspondence with its capacitance C. The larger the capacitance is, the larger the charge the capacitor can hold for a unit of voltage. Capacitance is defined as electrical charge over voltage: $C = \frac{Q}{u}$, giving $Q = u \cdot C$. The definition of electrical current I is a change of charge in respect to time: $I = \frac{dQ}{dt}$. Therefore:

$$I_C(t) = C \cdot \frac{du(t)}{dt} \tag{5.2}$$

Current driven through a resistor behaves according to Ohm's law: $u = R \cdot I$. In our case: $u(t) - u_{rest} = R \cdot I_R(t)$, where $u(t)$ is the voltage across the membrane and u_{rest} is the neuron's resting potential. Therefore:

$$I_R(t) = \frac{1}{R}(u(t) - u_{rest}) \tag{5.3}$$

Therefore, from Eq. 5.1 we conclude:

$$I(t) = \frac{1}{R}(u(t) - u_{rest}) + C \cdot \frac{du(t)}{dt} \tag{5.4}$$

Rearranging Eq.5.4 gives:

$$R \cdot C \cdot \frac{du(t)}{dt} = -(u(t) - u_{rest}) + R \cdot I(t) \tag{5.5}$$

Substitute $R \cdot C$ for τ (unit of time) and $(u(t) - u_{rest})$ for $V(t)$, gives:

$$\tau \frac{d}{dt} V(t) = -V(t) + R \cdot I(t) \tag{5.6}$$

This, by keeping in mind that $\frac{dV(t)}{dt} = \frac{du}{dt} - \frac{du_{rest}}{dt} = \frac{du}{dt}$, since $\frac{du_{rest}}{dt} = 0$ (u_{rest} is a constant).

The value of τ is of particular interest to dynamic modeling, commonly referred to as the *membrane time constant.* As we will see in the analysis below, τ is the time for the membrane potential to rise from its resting value to 63 % of its saturated value. Eq. 5.6 can be solved for different dynamic of $I(t)$. This will be the topic of discussion in the next few sections. Importantly, the LIF model is built upon the passive membrane model. When a threshold for spike initiation is reached, a spike is generated. The membrane voltage is then driven back to its resting potential where it stays for a predefined refractory period.

5.1.1 Membrane voltage for various input patterns

Here, we will briefly explore various solutions to Eq. 5.6:

- Flat input. The simplest case in which the input current I is 0 and the membrane voltage drops to its resting state
- Step current input. Demonstrating the model response to an injected input current I rising from 0 to some constant value
- Pulse input. Demonstrating the model response to both injection and termination of an input current I
- An arbitrary input

A detailed analysis of these solutions is available in [100].

5.1.2 Flat input

For a flat input, $I(t) = 0$ and Eq. 5.6 is reduced to:

$$\tau \frac{dV(t)}{dt} = -V(t). \tag{5.7}$$

Separation of variables gives:

$$\frac{1}{V(t)} dV(t) = -\frac{1}{\tau} dt. \tag{5.8}$$

We integrate both sides:

$$\int \frac{1}{V(t)} dV(t) = \int -\frac{1}{\tau} dt, \tag{5.9}$$

,

to receive:

$$\begin{aligned} ln|V(t)| &= -\frac{t}{\tau} + D \\ V(t) &= e^{-\frac{t}{\tau}} \cdot D \end{aligned} \tag{5.10}$$

Given a boundary condition in which $V(0) = V_0$, we derive: $e^0 \cdot D = V_0$, and therefore $D = V_0$. Substitution gives $V(t) = V_0 \cdot e^{-\frac{t}{\tau}}$ or:

$$u(t) = u_{rest} + V_0 \cdot e^{-\frac{t}{\tau}} \tag{5.11}$$

Eq. 5.11 expresses an *exponential decay* - an abundant solution in many biological-physical systems. The importance of τ to this exponent dynamic will be further discussed in the next solution.

5.1.3 Step current input

For a step current input I_0 at time t_0, Eq. 5.6 is resolved to:

$$u(t) = u_{rest} + R \cdot I_0 \left[1 - e^{-\frac{t-t_0}{\tau}}\right] \tag{5.12}$$

Lets verify Eq. 5.12. Its derivative is:

$$\frac{du(t)}{dt} = R \cdot I_0 \cdot \frac{1}{\tau} \cdot e^{\frac{-t}{\tau}}. \tag{5.13}$$

Note that the derivative of $e^{f(x)}$ is $f'(x) \cdot e^{f(x)}$. We can write Eq. 5.6 as:

$$u(t) - u_{rest} = R \cdot I_0 - \tau \cdot \frac{du(t)}{dt}. \tag{5.14}$$

Substitution gives:

$$u(t) - u_{rest} = R \cdot I_0 - \tau \cdot R \cdot I_0 \cdot \frac{1}{\tau} \cdot e^{\frac{-t}{\tau}} = R \cdot I_0 - R \cdot I_0 \cdot e^{\frac{-t}{\tau}}. \tag{5.15}$$

Assigning this result in the derivative of Eq. 5.12 gives:

$$\frac{du(t)}{dt} = \frac{1}{\tau} \cdot (u(t) - u_{rest} - R \cdot I_0), \tag{5.16}$$

which is equals to Eq. 5.6, verifying correctness of Eq. 5.12.

To better understand this solution, we will evaluate it for three time points: when t approaches infinity, when $t = t_0$ and when $t = \tau$.

- For a large value of t, $e^{-\frac{t}{\tau}}$ approaches 0, yielding:

$$u(t_\infty) = u_{rest} + R \cdot I_0 \tag{5.17}$$

- When $t = t_0$ (ahead of current injection), $e^{-\frac{t}{\tau}}$ approaches 1, yielding:

$$u(t_0) = u_{rest}. \tag{5.18}$$

  Between t_0 and t_∞, there is an exponential increase of voltage, rising from $u(t_0)$ to $u(t_\infty)$.

- When $t - t_0 = \tau$:

$$u(\tau) = u_{rest} + R \cdot I_0 \left[1 - e^{-\frac{\tau}{\tau}}\right]$$
$$u(\tau) = u_{rest} + R \cdot I_0 \left[1 - \frac{1}{e}\right] = u_{rest} + 0.63R \cdot I_0. \tag{5.19}$$

  τ is therefore the time it took for the membrane potential to reach 63 % of its saturated value.

One particularly useful form of Eq. 5.12 is achieved when it is used to define neuron activity a or the neuron's firing rate as a function of current I_0. Spiking rate a can be defined using:

$$a = \frac{1}{t_{ref} + t_{th}} \tag{5.20}$$

where t_{ref} is the refractory time period and t_{th} is the time to threshold. Since the spike period is small in comparison to the input integration phase, $t_{ref} + t_{th}$ constitutes the spike time, and therefore its inverse is the spike rate (frequency f is the inverse of cycle time t since $f = 1/t$)). In Eq. 5.12, we can substitute $u(t)$ for u_{th} (the voltage for which a spike is induced), $t - t_0$ for t_{th} and set u_{rest} to 0, yielding:

$$u_{th} = R \cdot I_0 \left[1 - e^{-\frac{t_{th}}{\tau}}\right] \tag{5.21}$$

Solving it for t_{th} would give:

$$t_{th} = -\tau ln(1 - u_{th}/R \cdot I_o) \tag{5.22}$$

By defining t_{ref} as a parameter, we can describe a with:

$$a = \frac{1}{t_{ref} - \tau ln(1 - u_{th}/R \cdot I_o)} \tag{5.23}$$

Eq. 5.23 would be particularly useful for neuromorphic representation, as we will see in Section 12.1.1.

5.1.4 Numerical modeling of pulse input

Eq. 5.12 can be enumerated by defining a numerical limit δt. We can iteratively calculate in each step i the contribution of the upcoming step ($u(i+1)$) to the voltage gained at the step before ($u(i)$). This can be done by defining u_∞ as the value that the membrane voltage exponentially decays towards at this time step:

$$u_\infty(i) = u_{rest} + R \cdot I(i), \tag{5.24}$$

thus yielding:

$$u(i+1) = u_\infty + (u(i) - u_\infty) \cdot e^{-\frac{dt}{\tau}} \tag{5.25}$$

This approach is particularly useful for computer simulations.

We can further analyze the pulse input case for a short current pulse I_0, initiated in t_0 for a period of Δ. Particularly, we would like to calculate the voltage $u(t_0 + \Delta)$. The solution to Eq. 5.6, which was given in Eq. 5.12, is also relevant here. At Δ, Eq. 5.6 is resolved as:

$$u(\Delta) = u_{rest} + R \cdot I_0 \left[1 - e^{-\frac{\Delta}{\tau}}\right]. \tag{5.26}$$

Since the Taylor expansion of e^x is $\sum_{n=0}^{\infty} \frac{x^n}{n} = 1 + x + ...$, by taking into account only $1 + x$ for e^x, we can approximate solution 5.6 to:

$$u(\Delta) = u_{rest} + R \cdot I_0 \cdot \frac{\Delta}{\tau} \tag{5.27}$$

Since $\tau = R \cdot C$, we can further simplify Eq. 5.27 to:

$$u(\Delta) = u_{rest} + I_0 \cdot \frac{\Delta}{C} \tag{5.28}$$

Since $I_0 \cdot \Delta$ is the charge deposited by the pulse, we can further simplify the solution to:

$$u(\Delta) = u_{rest} + \frac{Q}{C}. \tag{5.29}$$

An interesting analysis of this case study would be to consider the model's behavior for an instantaneous input pulse. Note that $u(\Delta)$ depends on the charge across the capacitor. Therefore, shortening the pulse by half while increasing its current level by two will result in the same voltage level, gained faster at $\Delta/2$. Therefore, if we consider an infinitely small pulse (Dirac function, $I(t) = Q \cdot \delta(\Delta)$), we will see an instantaneous rise of the membrane potential to $\frac{Q}{C}$. It will be followed by an exponential decay, where:

$$u(t) = u_{rest} + \frac{Q}{C} \cdot e^{-\frac{t}{\tau}} \tag{5.30}$$

5.1.5 Arbitrary input

Consider a series of pulses, each delivering a charge Q to the circuit. Their contributions to the membrane voltage add up linearly. At time t, the membrane voltage will be the sum contribution of all preceding pulses. This can be formalized as:

$$u(t) = u_{rest} + \int_{-\infty}^{t} \frac{1}{C} \cdot e^{\frac{t-t'}{\tau}} \cdot I(t')dt' \tag{5.31}$$

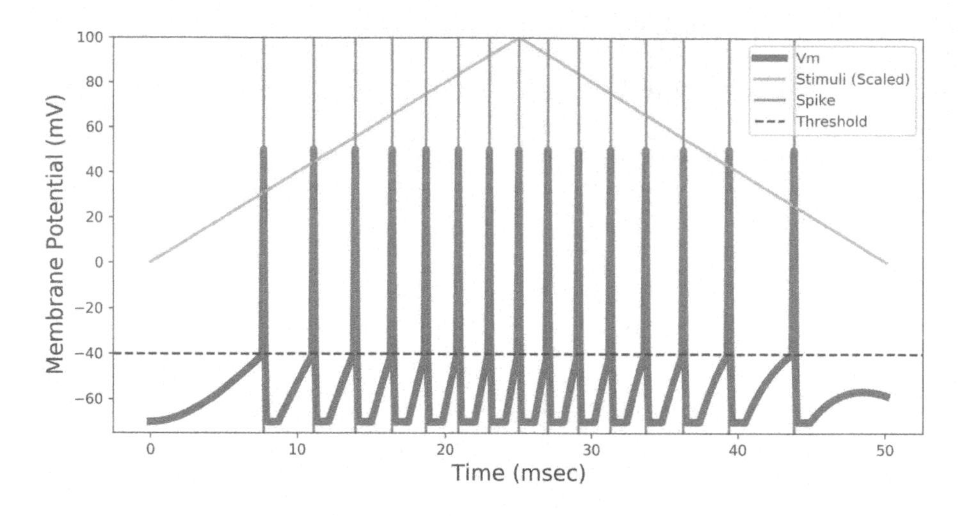

Figure 5.2 Simulation of Leaky integrate and fire model. The input current is colored in orange and the voltage membrane in blue. The voltage membrane rises exponentially fast until the threshold is reached. When the threshold is reached, a spike is generated and the membrane voltage is driven back to its resting potential, where it stays for a predefined refractory period.

Eq. 5.31 integrates the contribution of each pulse at time t' to the membrane potential at time t.

Simulated LIF is shown in **Figure 5.2**. Implementation is given in Python on the book website.

5.2 THE IZHIKEVICH NEURON MODEL

The Izhikevich neuron model is a quadratic LIF model with a recovery variable [141]. It is able to replicate several characteristics of biological neurons while remaining computationally efficient. The model is based on two differential equations:

$$\begin{aligned} \frac{du}{dt} &= 0.04u^2 + 5u + 140 - v + I \\ \frac{dv}{dt} &= a(b \cdot u - v) \end{aligned} \tag{5.32}$$

As with the LIF model, it requires manual resetting: if $u >= 30$mv, then $u = c$ and $v = v + d$.

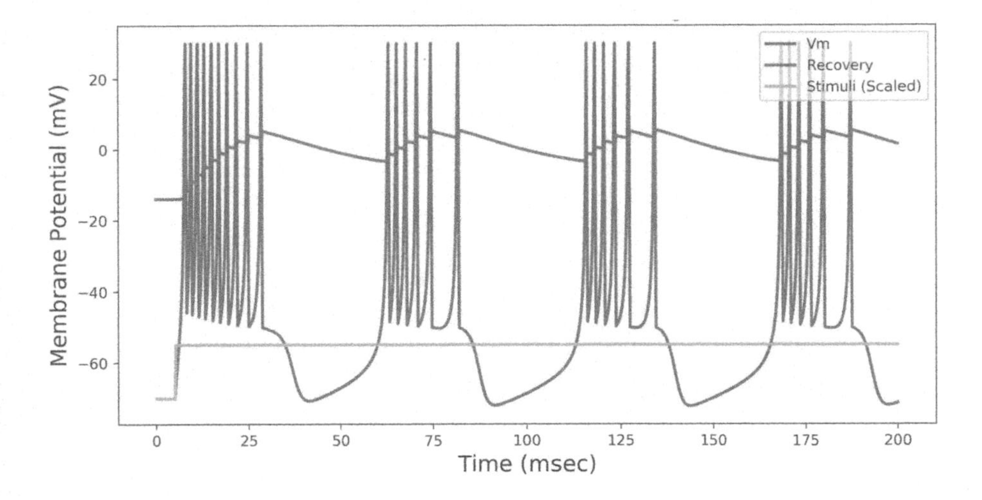

Figure 5.3 Izhikevich chattering spiking model. Chattering spiking mode is achieved when $a = 0.02$, $b = 0.2$, $c = -50$, and $d = 2$.

In this model, u represents the neuron's membrane potential, and v represents a membrane recovery variable which accounts for the activation of K^+ and inactivation of N_a^+ ionic currents. In contrast to LIF, parts of the Izhikevich neuron model were designed to fit observable biological dynamics. For example, the expression $0.04u^2 + 5u + 140$ was obtained by mathematically fitting the cortical neuron's spike initiation dynamic. The model has four parameters:

- Parameter a describes the time scale for membrane recovery
- Parameter b describes the membrane recovery's sensitivity to fluctuations in membrane potential, allowing the achievement of subthreshold spiking
- Parameter c describes the post-spike membrane potential
- Parameter d describes the membrane post-spike recovery

Different value combination of these parameters account for different neuronal dynamics, such as chattering (spike bursts) (**Figure 5.3**), fast spiking (**Figure 5.4**), and regular spiking (**Figure 5.5**). Model implementation is given in Python on the book website. The Izhikevich model is frequently used in large scale simulations, such as the ones performed by IBM (see Section 1.2.2).

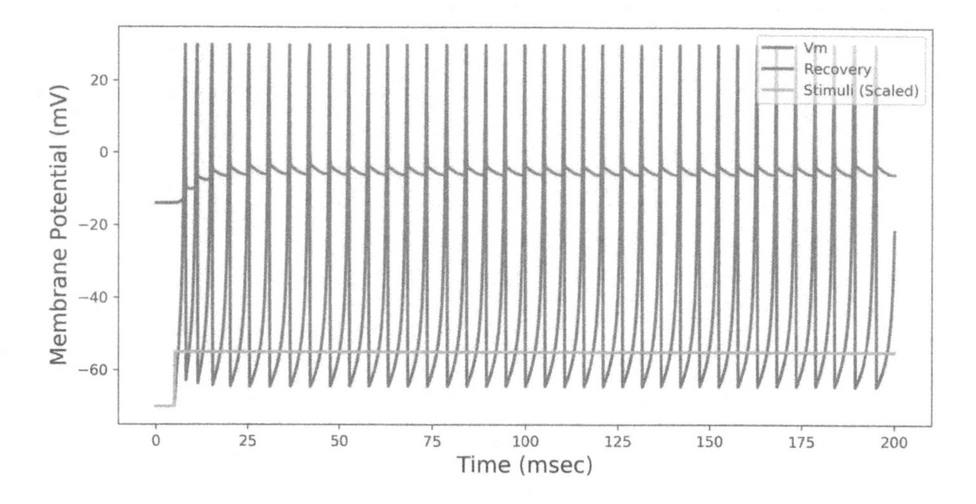

Figure 5.4 Izhikevich fast spiking model. Fast spiking mode is achieved when $a = 0.1$, $b = 0.2$, $c = -65$, and $d = 2$.

Note that the notations proposed here are slightly different from those proposed in Izhikevich's original paper in which the membrane voltage was annotated with v. Here, I changed the notations to keep the formulations consistent with the other models.

Both LIF and Izhikevich models have explicitly defined thresholds. They are therefore limited in their biological plausibility. The HH model is a realistic formulation in which spikes emerged from the complexity of its differential nature.

5.3 THE HODGKIN-HUXLEY MODEL

The HH model is the most famous and influential biological plausible neuron model. It models ion channels' conductance in terms of activation and inactivation variables [126]. Similar to the LIF model, in the HH model, capacitance C and conductance g $(= \frac{1}{R})$ represent the membrane's ability to separate ions and to drive them through the membrane. Ion's reversal potential is represented by a power source E. In the HH model however, two additional voltage-dependent channels are added, one modeling the conductance to N_a^+ and the other to K^+ (**Figure 5.6**).

The voltage-dependability and time-constant of the N_a^+ and K^+ channels are represented by differential equations. Each equation describes a dimensionless gate status using three gate variables: m, h, and

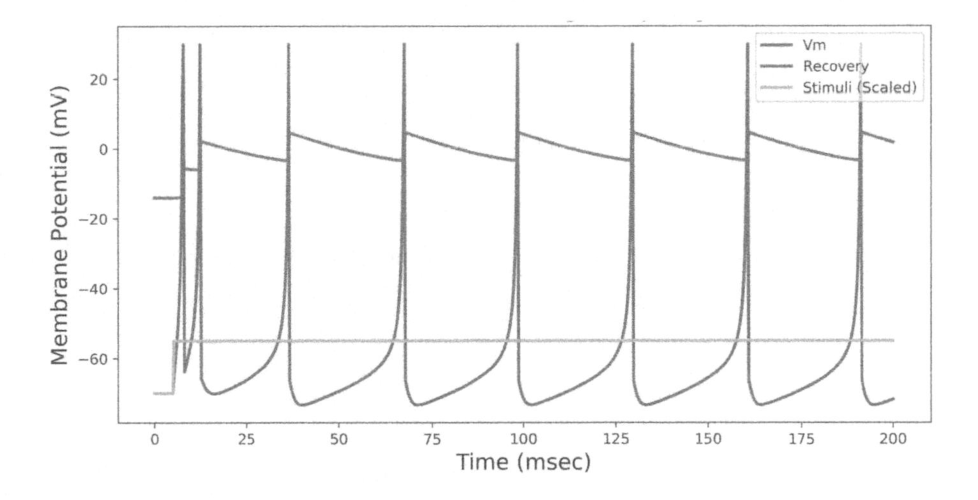

Figure 5.5 Izhikevich regular spiking model. Regular spiking mode is achieved when $a = 0.02$, $b = 0.2$, $c = -65$, and $d = 8$.

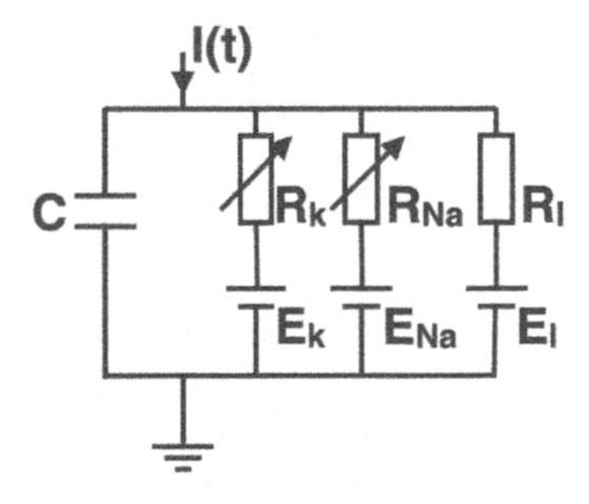

Figure 5.6 The Hodgkin-Huxley circuit model. The cell membrane is represented by a capacitor, and voltage-gated and leak ion channels. The electrochemical gradients driving the flow of ions are represented by batteries (E). The model takes into account N_a^+, K^+ ion channels and a leakage channel.

n (values between 0 and 1). While n defines the model's conductance to K^+, m and h define its conductance to N_a^+. The HH model is defined with:

$$I = C_m \frac{dV_m}{dt} + g_k n^4 (V_m - V_k) + g_{Na} m^3 h (V_m - V_{Va}) + g_l (V_m - V_l) + I(t) \tag{5.33}$$

where $I(t)$ is the input current. The values of m, h, and n are defined by two rate constants: α and β which are functions of V_m. The dynamic of n, m and h are defined with:

$$\begin{aligned} \frac{dn}{dt} &= \alpha_n(V_m)(1-n) - \beta_n(V_m)n \\ \frac{dm}{dt} &= \alpha_m(V_m)(1-m) - \beta_m(V_m)m \\ \frac{dh}{dt} &= \alpha_h(V_m)(1-h) - \beta_h(V_m)h \end{aligned} \tag{5.34}$$

where rates α and β are defined with:

$$\begin{aligned} \alpha_n(V_M) &= \frac{0.01(10 - V_m)}{e^{\frac{10-V_m}{10}} - 1} & \beta_n(V_m) &= 0.125 \cdot e^{\frac{-V_m}{80}} \\ \alpha_m(V_M) &= \frac{0.1(25 - V_m)}{e^{\frac{25-V_m}{10}} - 1} & \beta_m(V_m) &= 4 \cdot e^{\frac{-V_m}{18}} \\ \alpha_h(V_m) &= 0.07 \cdot e^{\frac{-V_m}{20}} & \beta_h(V_m) &= \frac{1}{e^{\frac{30-V_m}{10}} + 1} \end{aligned} \tag{5.35}$$

The resulting behaviors of m, h, and n in response to a step function are shown in **Figure 5.7** and the resulting ion currents are shown in **Figure 5.8**. In response to these ion currents, the membrane voltage exhibits emergent spikes, as shown in **Figure 5.9**. Note that this model does not account for the experimentally observed stochastic response of neurons to current injection [208]. Further expressions, which generalize the HH model allow easy transfer of electrophysiological measurements to the HH-parameter space were proposed, but they are outside the scope of this book.

While both LIF and the Izhikevich models only aim for a correct phenomenology of spiking, the HH model is "bio-physically meaningful" in the sense that it attempts to follow real biophysics from which spikes "naturally" emerged.

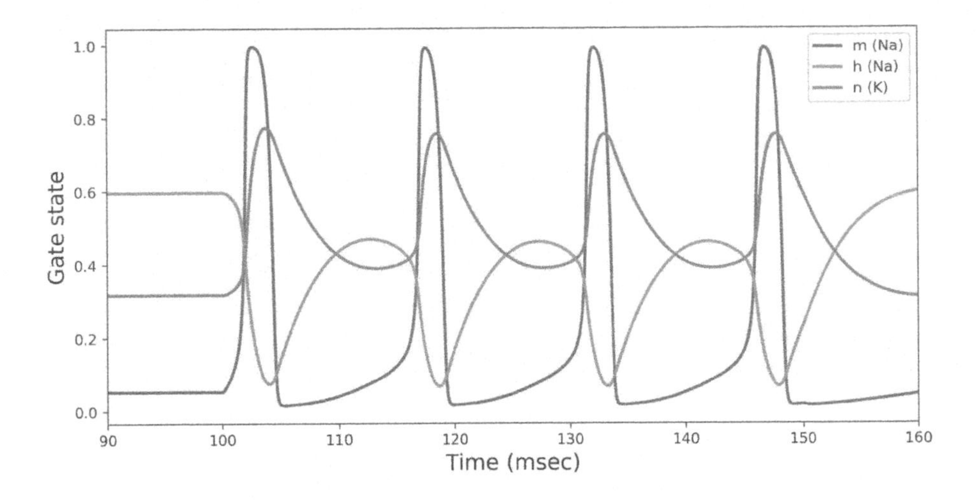

Figure 5.7 Gates dynamic in the Hodgkin-Huxley model. Each gate has its own dynamic dictated by Eq. 5.34. For example, while N_a^+ channel responds rapidly, K^+ channels has a slower dynamic, contributing to the spike waveform.

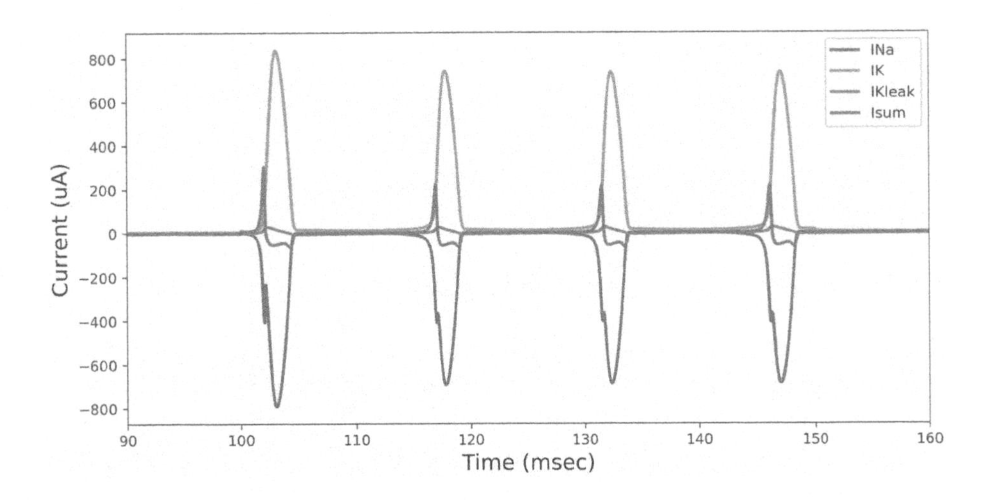

Figure 5.8 Currents dynamic in the Hodgkin-Huxley model. Ion currents are driven by the electrochemical forces emulated by the model and the state of the gates. N_a^+ and K^+ currents opposite each other, as N_a^+ is driven into the cell and K^+ is driven out of it. The total current dictates the formation of the action potential.

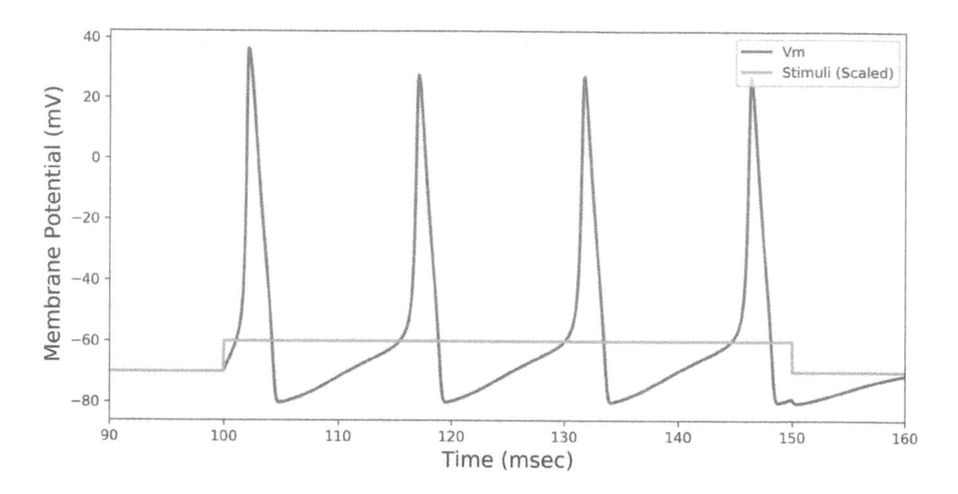

Figure 5.9 Spiking dynamic in the HH model. Spikes emerge from the model's governing dynamics, exhibiting a biologically plausible behavior.

While the three models described above were stimulated with explicitly injected input current I, real neurons receive spikes, arriving at their synapses.

5.4 SYNAPSE MODELING

Essentially, the synapse is just another channel which can replace the model's current term $I(t)$. A well-rounded review for synaptic modeling is available in [247]. One important formulation of synaptic modeling for the HH model, referring to Equation 5.33, is:

$$I(t) = I^{syn}(t) = g_{syn}(t)(u - E_{syn}) \tag{5.36}$$

where g_{syn} is the synapse's conductance and E_{syn} is its reversal potential, according to which the synapse is either excitatory or inhibitory. For example, when $E_{syn} = 55$ mV (similarly to E_{N_a}) the synapse will be excitatory and when $E = -75$ mV (similarly to E_K), it will be inhibitory. One simple formulation for g_{syn} would take the form of an exponential decay which we have encountered several times before:

$$g_{syn}(t) = \hat{g_{syn}} \cdot e^{\frac{-(t-t_0)}{\tau}} \tag{5.37}$$

This model assumes an instantaneous rise of the synaptic conductance $g_{syn}(t)$ from 0 to $\hat{g_{syn}}$ at time t_0, followed by an exponential decay with a time constant τ. Note that for $t < 0$, $g_{syn} = 0$. What would be the value of $\hat{g_{syn}}$? In Section 4.1.4, we described a synapse's dynamic as dependent on the availability of synaptic vesicles which are loaded with neurotransmitters. We can, therefore, define $\hat{g_{syn}}$ as a function of vesicles availability:

$$\hat{g_{syn}} = C \cdot p_{rel} \tag{5.38}$$

where C is a constant and p_{rel} is the percentage of available neurotransmitters.

Synapses are not constant entities as they change over time; they have an adjustable efficacy or weight. The weight of a synapse is affected by the density of the neurotransmitter receptors on the post-synaptic cell, vesicle density, and vesicle fusion rate.

5.5 SIMULATING POINT NEURONS

The three models discussed in the previous sections represent point processes – neurons with no spatial structure or morphology. However, they are powerful abstractions for neuronal activity, as they can manifest various neuronal features founds in biology. Biological plausibility usually comes at the expense of computability and scalability. This will be the next topic of discussion.

5.5.1 Biological plausibility vs. computational resources

In another seminal work from 2004, *Eugene Izhikevich* identified the different biological features each abstraction can provide, along with an estimate of the required computational resources in terms of FLOPS [142]. While the LIF model requires modest computational resources with only 5 FLOPS needed for evaluation, it can only express three spiking modalities: regular spiking, increased spike frequency with increased stimuli intensity, and voltage integration. However, the HH model is at the other extreme, providing 19 features, and requiring 1200 FLOPS. The Izhikevich model has the unique advantage of providing 21 features at the cost of only 13 FLOPS (**Figure 5.10**). The HH model, however, as opposed to the Izhikevich model, is biologically plausible.

5.5.2 Large scale simulations of point processes

This book is not aimed at giving a precise map of large scale simulations, as progress in simulation scale and plausibility is constantly increasing.

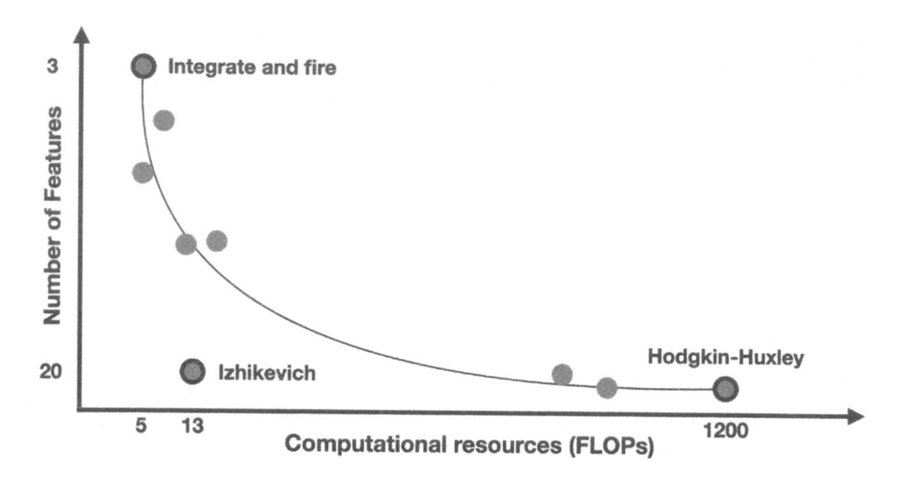

Figure 5.10 Comparison of the number of neuronal features (e.g. spike frequency adaptation and phasic spiking) and the required computational resources (needed to simulate the model for 1ms) for the LIF, HH and the Izhikevich models. Data from [142].

However, it might be useful to grasp the vast numbers large-scale simulations can entail. In 2007, *Eugene Izhikevich* demonstrated the scalability of his neural model (Section 5.2) by executing a simplified thalamocortical simulation comprised of 10^6 neurons which were categorized into 22 sub-types and connected by $\approx 5 \cdot 10^8$ synapses [144]. In 2013, IBM used their supercomputers to simulate $65 \cdot 10^9$ Izhikevich neurons with $16 \cdot 10^{12}$ synapses [233]. In this work, IBM researchers followed the long-distance wiring diagram of the macaque brain which spans the cortex, thalamus, and basal ganglia. In a later work, IBM simulated 10^9 neurons and 10^{13} synapses (the scale of the cortex of a cat), following a wiring schematic inspired by the thalamocortical system [15]. In 2019, following ten years of research and development, IBM's TrueNorth chip climaxed with the NS16 TrueNorth system which supports the simulation of 10^8 spiking neurons with neuromorphic hardware [72]. Simulating these networks, despite being low on biological details, requires tremendous amounts of computing resources for which neuromorphic hardware are becoming applicable.

One prominent neuromorphic board is the SpiNNaker, designed at the University of Manchester. The SpiNNaker was shown to be able to support LIF-based cortical simulations with similar accuracy as traditional supercomputers. For example, it was used to simulate $80 \cdot 10^3$ neurons with $900 \cdot 10^6$ synapses [9]. However, the SpiNNaker was not designed to support the time scales of biological systems and it was therefore configured to slow down. This cortical simulation was performed on a SpiNNaker system with 5 SpiNNaker boards, each with $48 \cdot 18$ cores, constituting a total of 288 chips and 5,174 ARM cores. For comparison, the simulation was also performed on 64 2.5 GHz Intel Xeon E5- 2680v3 processors. Surprisingly, results show that both systems have a comparable energy consumption per synaptic event. While energy consumption per synaptic event for the supercomputer was measured at $5.8 \cdot 10^{-6}$ J, for the SpiNNaker, it was $5.9 \cdot 10^{-6}$ J. However, non-biologically-tailored SpiNNaker simulations had energy consumption of $8 \cdot 10^{-9}$ J per synaptic event. This reduction in efficiency was mainly due to mismatched timescales of the optimized operation point. In another study, a GPU-based simulation of the same neuronal circuit had higher energy efficiency with the consumption of 10^{-6} J per synapse event and faster time to solution [158]. A year after that, the race was on. A simulation of the same circuit was demonstrated on the SpiNNaker board, featuring GPU energy performance in real-time (GPU implementation was $2x$ real-time, and supercomputer implementation was $3x$) [244].

As architectures and algorithms continue to develop, there is still a long way to go for achieving brain-level performance in which the estimated power consumption is only 10^{-15} J per synaptic event (Section 4.2). These results highlight the need for further optimizations of neuromorphic hardware for biologically detailed simulations.

5.6 CASE STUDY: A SNN FOR PERCEPTUAL FILLING-IN

From the scientist's perspective, neuromorphic computing holds promise for computational models with high explanatory power. Computational biologically plausible models connect high cognitive qualities (an emergent property) to low-level circuitry (SNN). In this case study, we will review two biologically plausible "point" spiking neural network models for perceptual filling-in [60].

5.6.1 Perceptual filling-in

Visual perception is initiated with low-level processing in the retina, from which it is propagated to the Lateral Geniculate Nucleus (LGN) and the primary visual cortex (V1). While in V1, visual data represent

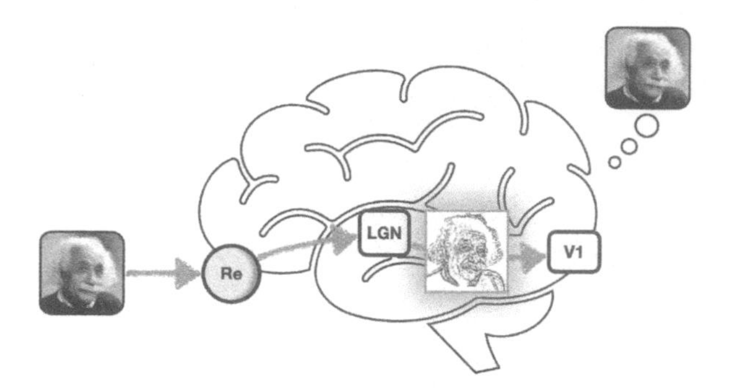

Figure 5.11 Visual perception is initiated in the retina (Re), from which it is propagated to the LGN and the primary visual cortex (V1). While in V1, visual data represent Spatio-temporal edges; the perceived image has complete filled-in surfaces.

Spatio-temporal edges [193]; the perceived image has complete filled-in surfaces. Therefore, the brain reconstructs visual constructs from their edges [121] (**Figure 5.11**).

Numerous visual phenomena (or illusions) involve a perceptual filling-in [160]. Among them are:

- **The watercolor illusion** [228]. White areas are perceived as brightly colored with the color of the polygon surrounding them, as long as the polygon is itself surrounded by a thin, darker border. This is more pronounced when the polygon's inner and outer borders are of complementary colors (e.g., orange and purple).
- **The neon color spreading** [290]. Colored borders are perceived between the edges of a colored object and the background in the presence of black lines.
- **Afterimage filling-in** [174] (or, the color dove illusion [25]). The disappearance of a colored background causes an empty shape to obtain the hue of the disappeared background.
- **Filling-in in the blind spot** [237]. The brain fills in the visual information in the retinal blind spot (created due to the optic nerve ending) from the surrounding picture; thus, we do not perceive the existence of such a spot.

Extensive empirical research on these visual illusions has led to two prominent theories governing perceptual filling-in [160]:

- **Symbolic or cognitive theory**. Low-level visual areas represent the contrast information at the surface edges, and the color and shape of the surface are described as metadata in higher areas.
- **Isomorphic theory**. An activation pattern spreads across the visual cortex's retinotopic map, from the edges to the interior of the surface, creating a perceptual surface.

The underlying mechanism of perceptual filling-in remains unclear, as there is experimental evidence supporting both hypotheses.

5.6.2 Mathematical formulation

In a computational model proposed by *Hadar Cohen-Duwek* and colleagues [61], it was shown that perceptual filling-in could be described using a Poisson equation. This mathematical formulation can reconstruct an image from its gradients.

A relevant starting point would be the diffusion, or the heat equation:

$$\frac{dI_p}{dt} - \Delta I_p(x, y) = div(\nabla I_s) \tag{5.39}$$

where I_p is the perceived (or the reconstructed) image, I_s is the input image, ∇ is the gradient operator, defined with: $[\frac{d}{dx}, \frac{d}{dy}]$, Δ is the Laplacian operator, defined with: $[\frac{d^2}{dx^2}, \frac{d^2}{dy^2}]$, and div is the divergence operator. In accordance with previous reports, the perceived image is reconstructed almost instantaneously, and is referred to as "immediate filling-in" [174]. Considering this fast dynamic, the dynamic phase of the diffusion equation can be ignored and thus reduced to the steady-state Poisson equation:

$$\Delta I_p(x, y) = -div(\nabla I_s) \tag{5.40}$$

While the model describes a mathematical formulation for perceptual filling-in, it does not imply how and where it is realized with the visual system's neural activity. In this case study, we will explore two neuronal implementations of this mathematical model using SNNs. These implementations support the isomorphic theory and demonstrate how a biologically plausible neuronal model can lead to a perceived reconstruction of an image from its gradients. Particularly, we solve the Poisson equation by two distinct networks:

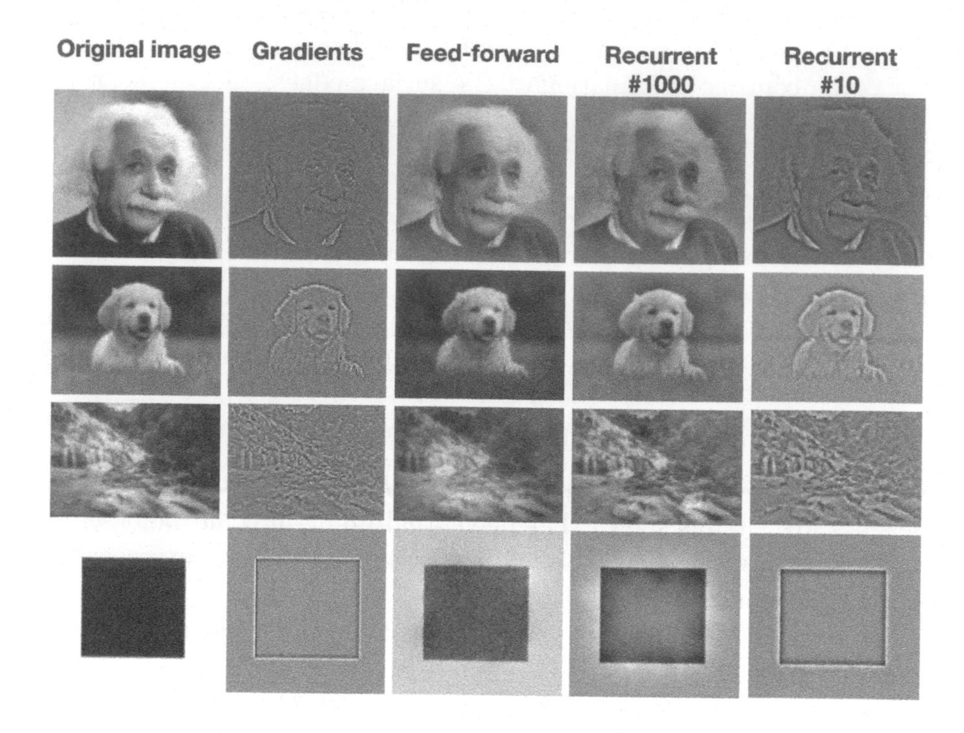

Figure 5.12 The original image and its Laplacian are in the 1st and 2nd columns, respectively. The results of the feed-forward and recurrent (at iteration 1000) networks are in the 3rd and 4th columns, respectively. The recurrent method result at iteration 10 is in the 5th column. Landscape image by BerryJ, Dog image by Hebrew Matio, Einstein photo by Miquel Perello Nieto, CC BY-SA 4.0.

- Feed-forward SNN in which a weight matrix was optimized to generate a solution
- Recurrent SNN which follows evidence-based feedback connections or horizontal connections

Models' simulation results are shown in **Figure 5.12**. Pixels were represented with spiking neurons following NEF, which will be discussed later in the book.

5.6.3 Feed-forward SNN for perceptual filling-in

The feed-forward SNN solves the Poisson equation by using matrix manipulations. The model schematic is shown in **Figure 5.13**.

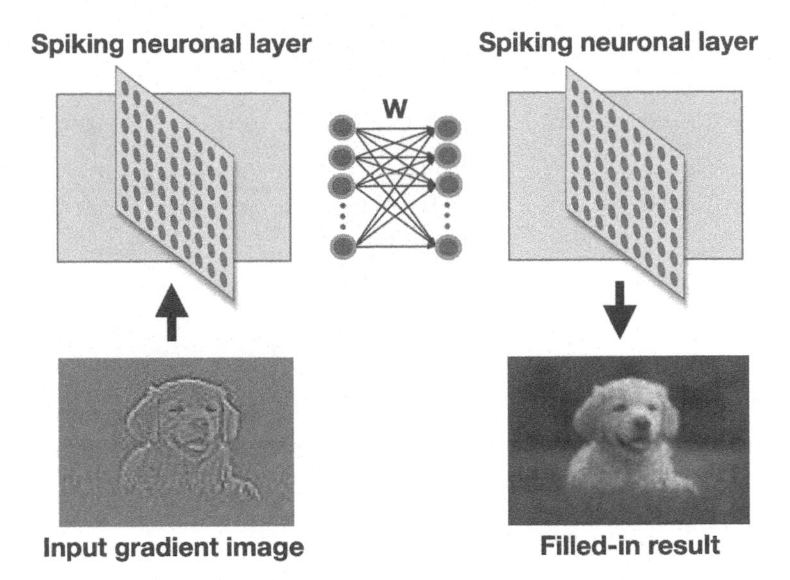

Figure 5.13 A feed-forward SNN for perceptual filling-in, where dense connections of two spiking layers reconstruct the input gradient image. Dog image by Hebrew Matio, CC BY-SA 4.0.

By using a finite difference numerical method [297], Equation 5.40 can be rewritten as a linear system:

$$Au = b \tag{5.41}$$

where u is the image to be reconstructed, u is a column vector representing the pixels of the image arranged in a natural ordering, and A is the Laplace matrix defined with:

$$A = \begin{bmatrix} D & -I & 0 & & 0 & 0 & 0 \\ -I & D & -I & \cdots & 0 & 0 & 0 \\ 0 & -I & D & & 0 & 0 & 0 \\ & \vdots & & \ddots & & \vdots & \\ 0 & 0 & 0 & & D & -I & 0 \\ 0 & 0 & 0 & \cdots & -I & D & -I \\ 0 & 0 & 0 & & 0 & -I & D \end{bmatrix} \tag{5.42}$$

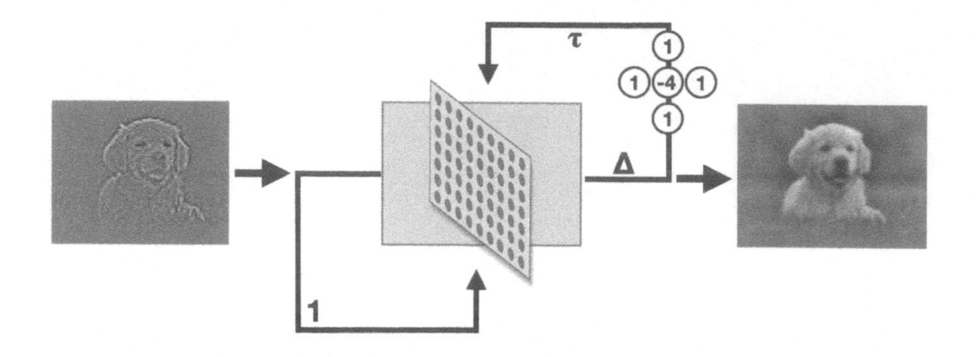

Figure 5.14 A recurrent SNN for perceptual filling-in, where an image is reconstructed from an input gradient image iteratively over time through recurrent (horizontal) connections. Dog image by Hebrew Matio, CC BY-SA 4.0.

where I is the identity matrix and D is given by:

$$D = \begin{bmatrix} 4 & -1 & 0 & & 0 & 0 & 0 \\ -1 & 4 & -1 & \cdots & 0 & 0 & 0 \\ 0 & -1 & 4 & & 0 & 0 & 0 \\ & \vdots & & \ddots & & \vdots & \\ 0 & 0 & 0 & & 4 & -1 & 0 \\ 0 & 0 & 0 & \cdots & -1 & 4 & -1 \\ 0 & 0 & 0 & & 0 & -1 & 4 \end{bmatrix} \tag{5.43}$$

Here, we assume that the visual system has a fixed W matrix, where $W = A^{-1}$, and the reconstructed image vector u can be calculated following Eq. 5.41 as $u = Wb$. We implemented this method in SNN by connecting two neuron ensemble layers with weight matrix W in an all-to-one connectivity scheme.

5.6.4 Recurrent SNN for perceptual filling-in

The recurrent method was implemented with a single layer, where feedback connections were defined from the layer to itself. This connectivity scheme can, therefore, be referred to as horizontal. The model schematic is shown in **Figure 5.14**.

By rearranging the dynamic form of Equation 5.39 as: $\frac{dI_p}{dt} = div(\nabla I_s) + \Delta I$, we can use Equation 12.36 (which will be discussed

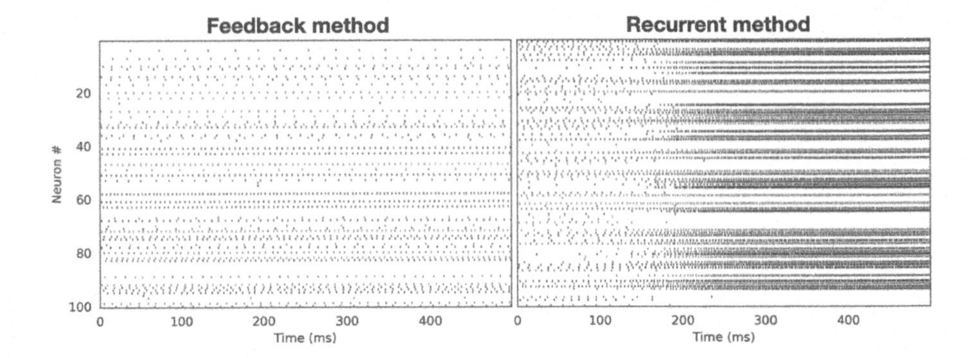

Figure 5.15 Raster plot for the black box image in the feedback (left) and the recurrent (right) methods. Spikes for the 100 neurons representing the central 5 pixels are shown (each pixel is represented by 20 neurons).

later), to define the feedback connection A for the realization of the Poisson equation as:

$$A = \tau(div(\nabla I_s) + \Delta I) + I \tag{5.44}$$

This system can be iteratively defined using:

$$I_k = \tau(div(\nabla I_s) + \Delta I_{k-1}) + I_{k-1} \tag{5.45}$$

The recurrent method iteratively reconstructs the perceived image I_k for each time step k.

Note that the recurrent method is not explicitly restricted to horizontal connections, as it can also be implemented with multiple neural layers. Thus, instead of horizontal connections, signals can be transmitted to a higher layer and then transferred back to the original layer with recurrent connections. In contrast to the feed-forward method, this method requires numerous iterations to converge and reconstruct the image. Notably, convergence is also apparent in the neuronal activity, as shown in the raster plots in **Figure 5.15**.

The absolute maximal changes across all pixels in the perceived image, over sequential iterations, are shown in **Figure 5.16**. Convergence is indicated when maximal change reduces to zero. To further realize the neuron number constraint, we monitored convergence, where the number of neurons representing a pixel was reduced from 20 to 10. It seems that with a small number of neurons, the solution diverges rather

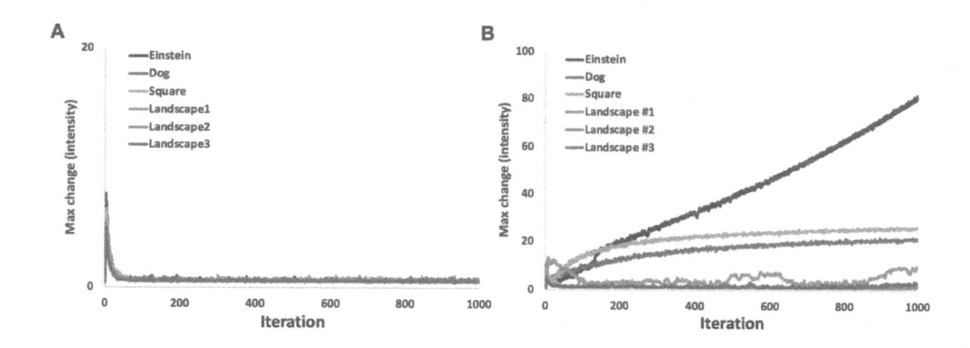

Figure 5.16 Convergence plot of the feedback method with 20 neurons per pixel (A) and with 10 neurons per pixel (B). The x-axis represents the number of iterations and the y-axis represents the maximum absolute difference between two sequential iterations. Only landscape 1 was shown in **Figure 5.12**.

than converges, except for the landscape image which features long and continuous edges on many small surfaces.

5.6.5 Is it biologically plausible?

In this case study, we introduced two biologically plausible computational methods which can serve as potential underlying filling-in neural mechanisms in the brain. Both methods were implemented using SNNs and were demonstrated with the reconstruction of an image from its Laplacian. Both methods are consistent with the isomorphic hypothesis since their resulted reconstruction was not directly stimulated by the input image. Although both approaches solve the same equation, their neural mechanisms are distinctive. While the recurrent method iteratively solves the Poisson equation using a horizontal connectivity scheme, the feed-forward method uses a weight matrix representing direct dense connectivity to do the same.

Several experimental findings support the spread of filling-in activities in the V1 area. For example, *Sang Wook Hong* and colleagues found that the response to a surface interior is delayed, relative to the response to the surface's edge, in a time constant proportional to the distance between a receptive field and the edge [129]. Our recursive method is consistent with this result, as the surface's interior is filled in at a later iteration than the area near the edges, thus suggesting that the recursive method may emulate the filling-in process in V1. As demonstrated by the black square example, filling-in is not complete for input images containing a large surface. This uncompleted filling-in phenomenon is consistent with experimental findings [315]. Note that the filling-in is completed for small surfaces or input with multiple edges (such as a natural image with many small surfaces). This implies that natural images (with many edges) can be fairly reconstructed at V1. This is also consistent with our findings, according to which the filling-in of natural landscapes is less prone to neuronal resources.

In contrast to the recurrent method, the feed-forward SNN is a non-iterative, direct method. Its filling-in performance is better than the recurrent method, as the hole in the center of the black square image example was filled. In terms of biological plausibility, this technique's drawback is that it entails all-to-one connections (dense matrix). This implies that neurons at a higher layer are connected to almost every neuron at a lower layer. A dense connectivity scheme is inconsistent with the receptive field organization, often a characteristic of neuronal visual layers [253]. This biologically implausible architecture might be resolved by separating the visual field into distinct regions. Independent visual fields of view can be reconstructed (from their local gradients) and "stitched" together at a higher visual layer.

Examining both approaches, while the recurrent method represents a slow spread of neural filling-in activities in V1, the feed-forward method represents fast neural filling-in activities in higher visual areas. It might be possible that both approaches are present in the brain.

5.7 GLOSSARY

Point neuron: A neuron model with no spatial structure or morphology.

The Hodgkin-Huxley model: The most famous and influential biological plausible neuron model. It models a neuron's ion channels' conductances in terms of activation and inactivation variables.

The Izhikevich neuron model: A quadratic LIF type model with a recovery variable, able to replicate several characteristics of biological neurons while remaining computationally efficient.

The leaky integrate and fire model: A model representing a neuron with a combination of a "leaky" resistor and a capacitor. The most abundant neuronal model for large scale simulations.

5.8 FURTHER READING

- One great resource for neuronal modeling is the book, "Neuronal Dynamics," by *Wulfram Gerstner* and colleagues [100] which is available as an online book at neuronaldynamics.epfl.ch.
- An insightful road map for brain modeling in which various guidelines and modeling efforts are described is given in [46].
- **Section 5.1**
  - A description of LIF from an historical perspective is given in [47].
  - An in-depth review of LIF is given in [48], along with a comparison with the HH neuron model and with electrophysiological data. A review on LIF, focusing on in-homogeneous synaptic input and network properties, is given in [48].
- **Section 5.2**
  - Dive into the works of *Eugene Izhikevich* by reading his publications on the Izhikevich neuron model in [141], on the exploration and comparison of neuronal models in [142] and on the utilization of the Izhikevich model in a large scale simulation in [144].
- **Section 5.3**
  - In his article: "What was Hodgkin and Huxley's Achievement?", *Arnon Levy* outlines the qualities of the HH model [170].
  - The ongoing impact of Hodgkin's and Huxley's ideas on neuroscience is given in [52].
- **Section 5.4**
  - A review of synapse modeling is given in [247].
  - A description of a synapse model that follows a dynamic stochastic process and which can be integrated into standard models of artificial neural networks is given in [185].

- **Section 5.5**
  - A short lecture by *Tadashi Yamazaki*, describing a model of $1mm^3$ cerebellar module on a computer, comprising $1 \cdot 10^6$ point processes, is availible at training.incf.org. This lecture is part of the course "Neuromorphic computing and challenges," offered by the online hub TrainingSpace.
- **Section 5.6**
  - The watercolor illusion is demonstrated at illusionsindex.org/i/watercolour-illusion
  - The neon color spreading is demonstrated at illusionsindex.org/i/neon-color-spreading
  - The afterimage filling-in or the color dove illusion is demonstrated at illusionsindex.org/ir/colour-dove-illusion

CHAPTER 6

Models of morphologically detailed neurons

Abstract

Real neurons have dendrites, axons, and synapses, all embedded in complicated densely wired structures. A neuron's complicated morphology is closely related to its functions and, therefore, capturing accurate morphologically and physiologically details are essential to link behavior to its underlying biological mechanisms. In this chapter, we will explore the rationale and some of the techniques for morphologically detailed neuronal modeling. We will particularly scale the dimensionless point processes we discussed earlier to dimensioned shapes using the cable equation and further scale them to branching morphologies using the compartmental model. Finally, we will use retinal directional selectivity as a case study.

6.1 WHY MORPHOLOGICALLY DETAILED MODELING?

Real neurons feature various shapes; they are often characterized by the different distribution of synapses and ion channels. Importantly, a neuron's unique features are closely related to its functions.

In his paper "Mapping function onto neuronal morphology," *Terrence Sejnowski* and colleagues iteratively modulated neurons' morphology with a genetic algorithm, such that they are optimized for linear summation or spike-order detection [269]. Neurons in the avian auditory brain stem implement linear summation by placing their synapses at two long and thin dendrites' ends. Due to the distance between them, their mutual influence on each other is minimized such that their contribution to neuron response is linear. In the problem of spike-order detection, the preferred neuron's morphology would have placed synapses with their dynamic aligned in a particular direction (e.g., left to right). Cortical pyramidal neurons achieve this by having two types of dendrites with significantly different diameters ($5x$), each carrying different synapses. By evolving dendrite morphology in the computer, artificial neurons were demonstrated to feature similar morphological characteristics. They thus showed the close link between a neuron's morphology to its functionality.

Another example is the Starburst Amacrine Cells (SACs) which have a fundamental role in retinal vision processing. SACs are important to many of the retina's visual processing: among them is direction sensitivity. SACs react differently to light coming from different directions. SAC's directionality is tightly connected to its morphology and the biophysical characteristics of its processes (in SAC, axons and dendrites are synonymous and are called processes). It was shown that non-symmetrical SACs have impaired directionality [214]. Responding with directional preference constitutes non-homogeneous characteristics. SACs, for example, feature proximal sustained synapses and distal transient synapses [109] which are carefully distributed, such that a distinct directional preference is maintained [292]. Moreover, SACs regulate their dendritic overlap with their neighboring SACs to ensure uniform coverage of the retinal field of view [153].

In most cases, the mapping between biological complexity and function is unknown. This fact makes simulation and modeling essential to our fundamental understanding of neuronal mechanisms. Computational models that abstract away such biological details will inherently fail to explain the functional significance of the intricacy of neuronal organization [192]. In this section, we will scale up point processes to support detailed morphological and dynamical modeling.

6.2 THE CABLE EQUATION

The origins of the cable theory dates back to 1855, when Lord *William Thomson Kelvin* provided the mathematical basis necessary for laying the transatlantic telegraph cable [127]. This section will briefly describe the mathematics behind the cable equation as it applied to neuronal modeling. An elaborated description is given in [100].

Advancing from point processes to detailed models requires the simulation of voltage propagation through space and time. To this aim, a cable or a continuous representation of a segment of a dendrite or an

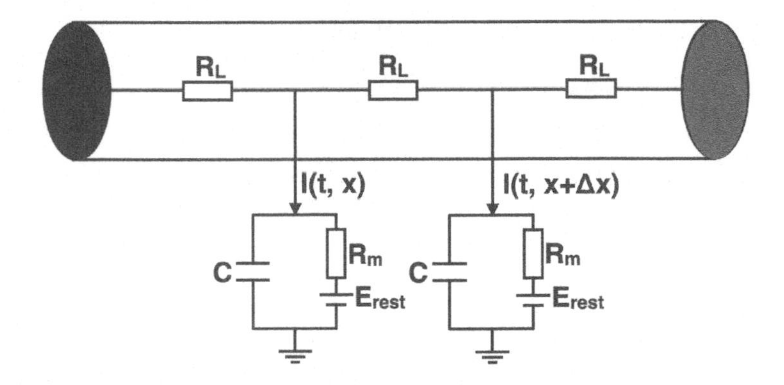

Figure 6.1 Schematic of a cable model. A cable discretized to interconnected passive LIF neuronal models.

axon can be discretized to be comprised of interconnected entities. Each entity provides the initial conditions for its neighboring entity. For example, a cable can be described using numerous connected LIF neuronal models, as demonstrated in **Figure 6.1**.

We will briefly analyze this model while generalizing it beyond LIF for it to be applicable to other conductance-based models, such as the HH model. To do so, we will define a conductance-based model to be the sum of currents driven through a capacitor C and other ion currents (e.g., I_{leak}, I_{N_a} and I_K):

$$I(t) = C \cdot \frac{du}{dt} + \sum_{ion} I_{ion}(t) \tag{6.1}$$

In the cable equation, we need to take both time and space into account. Referring to **Figure 6.1**, $I(t,x)$ is the current driven out of the cable and into the conductance model (here, LIF) in time t and position x. From Eq. 6.1 we derive:

$$I(t,x) = C \cdot \frac{d}{dt} u(t,x) + \sum_{ion} I_{ion}(t,x) \tag{6.2}$$

From the conservation of current, $I(t,x)$ equals the incoming current through resistor R_L driven from the previous section ($I_L(t,x)$) minus the output current $I_L(t,x+\delta x)$ (the current driven to the next section). Adding the possibility of an excitation current $I^{ext}(t,x)$, we conclude:

$$I(t,x) = I^{ext}(t,x) + I_L(t,x) - I_L(t,x+\delta x) \tag{6.3}$$

Considering Ohm's law, we can define:

$$I_L(t,x) = \frac{u(t,x-\delta x) - u(t,x)}{R_L}$$
$$I_L(t,x+\delta x) = \frac{u(t,x) - u(t,x+\delta x)}{R_L} \tag{6.4}$$

We can therefore reformulate Eq. 6.3 as:

$$I(t,x) = I^{ext}(t,x) + \frac{u(t,x-\delta x) - u(t,x)}{R_L} - \frac{u(t,x) - u(t,x+\delta x)}{R_L} \tag{6.5}$$

We can rearrange it and use Eq. 6.2 to conclude:

$$C \cdot \frac{d}{dt}u(t,x) + \sum_{ion} I_{ion}(t,x) = I^{ext}(t,x) + \frac{u(t,x-\delta x) - 2u(t,x) + u(t,x+\delta x)}{R_L} \tag{6.6}$$

Let's consider the cable's discretization. R_L is dependent on a discretization unit length dx. Therefore:

$$R_L = r_L \cdot \delta x \tag{6.7}$$

The same holds for C, I^{ext} and I_{ion}. Substituting these in Eq. 6.6, and rearranging the result, gives:

$$C \cdot \delta x \cdot \frac{d}{dt}u(t,x) + \sum_{ion} i_{ion}(t,x) \cdot \delta x$$
$$= i^{ext}(t,x) \cdot \delta x + \frac{u(t,x-\delta x) - 2u(t,x) + u(t,x+\delta x)}{r_L \cdot \delta x} \tag{6.8}$$

We can rearrange it and apply the limit $\delta x \to 0$ to conclude:

$$\lim_{\delta x \to 0} \frac{u(t,x-\delta x) - 2u(t,x) + u(t,x+\delta x)}{r_L \cdot (\delta x)^2}$$
$$= C \cdot \frac{d}{dt}u(t,x) + \sum_{ion} i_{ion}(t,x) - i^{ext}(t,x) \tag{6.9}$$

Solution to Eq. 6.9 gives the *cable equation*:

$$\frac{d^2}{dx^2}u(t,x) = C \cdot r_L \cdot \frac{d}{dt}u(t,x) +$$
$$r_L \cdot \sum_{ion} i_{ion}(t,x) - r_L \cdot i^{ext}(t,x) \tag{6.10}$$

Now, we can define $\sum_{ion} i_{ion}(t,x)$ to express the behavior of various neuronal "cables" such as passive dendrite or an axon.

6.2.1 Passive Dendrite

For a passive dendrite (LIF-based), $\sum_{ion} i_{ion}(t,x) = leak$. By referring to Eq. 5.4 we conclude:

$$\sum_{ion} i_{ion}(t,x) = \frac{u(t,x)}{R_m}. \tag{6.11}$$

By defining the terms: $\tau_m = C \cdot R_m$, and $\frac{R_m}{R_L} = \lambda^2$, and rearranging Eq 6.10, we get the *passive cable equation*:

$$\lambda^2 \frac{d^2}{dx^2} u(t,x) = \tau_m \frac{d}{dt} u(t,x) + u(t,x) - r_m \cdot i^{ext}(t,x) \tag{6.12}$$

6.2.2 Axon

The mathematical description of an axon is practically identical to that of a dendrite with active ion channels. Considering Eq. 6.10, we can remove the external current source i^{ext}, and add $R_L \cdot i_{ion}[u](t,x)$ to express active channels, yielding:

$$\frac{d^2}{dx^2} u(t,x) = c \cdot R_L \cdot \frac{d}{dt} u(t,x) + R_L \cdot \frac{u(t,x)}{r_m} - R_L \cdot i_{ion}[u](t,x) \tag{6.13}$$

The ionic currents can be described, as was mentioned in Section 5.3 by using the HH model:

$$i_{ion}[u](t,x) = g_{Na} m^3(t,x) h(t,x)(u(t,x) - E_{Na}) + g_k n^4(t,x)(u(t,x) - E_K) \tag{6.14}$$

6.2.3 Simulating the cable equation

We will use the NEURON simulation framework which was discussed in Section 3.2 to simulate the cable equation. The simulation code is provided on the book website.

Using the cable equation, we can trace changes in time ($\frac{d}{dt}$) and space ($\frac{d}{dx}$). For demonstration purposes, we will define a simple neuron comprised of a dendrite and a soma (stick and ball model; see NEURON's documentation for specific implementation details). Briefly, the cell soma is defined with a diameter of $12.6 \cdot 10^{-6}$ m, axial resistance of $100 Ohm \cdot cm$, and a membrane capacitance of $1 \frac{\mu F}{cm^2}$. Since we would like the soma to generate spikes, we will define it with a HH conductance

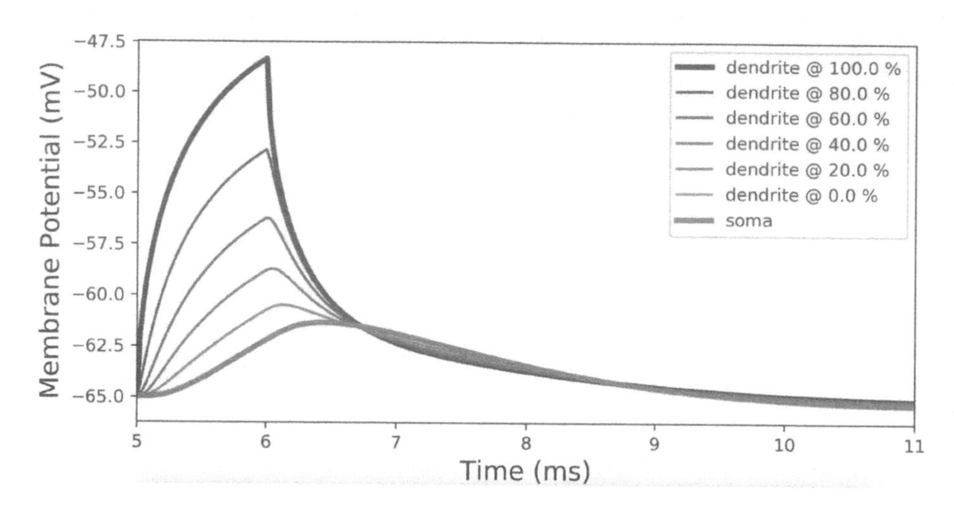

Figure 6.2 Simulating the cable equation with a sub-threshold stimulation. As we move further away from the stimulation point, voltage level decreases. Point of current injection was labeled as 100%. Voltage level at the soma is not sufficient to initiate a spike.

model, with $g_{N_a} = 0.12\frac{S}{cm^2}$, $g_K = 0.036\frac{S}{cm^2}$, and $g_l = 0.0003\frac{S}{cm^2}$. The dendrite will be connected to the soma, has a length of $200 \cdot 10^{-6}$ m, and a diameter of 10^{-6} m. Axial resistance and membrane capacitance were defined similarly to the soma. Since in this simple example, the dendrite passively propagates voltage, we can define it with passive LIF properties, where $R_m = 0.001\frac{S}{cm^2}$ and the reversal potential is -65 mV. To generate current stimulus $I(t)$, we will define a current clamp, connect it to the proximal part of the dendrite (at the opposite side of soma) and configure it to produce a 0.1 nA, 1 msec current pulse. We will measure voltage propagation by monitoring voltages at seven different locations across the dendrite simultaneously. Results show that voltage decays as it propagates down the dendrite to the soma (**Figure 6.2**).

To induce spike generation, the current pulse was magnified to 0.3 nA, reaching the voltage threshold in the soma (**Figure 6.3**).

In this simulation, the soma was defined with a single segment and the dendrite with 101 segments. How should we determine the appropriate number of segments and the extent to which the number of segments will change the results? When is it accurate enough?

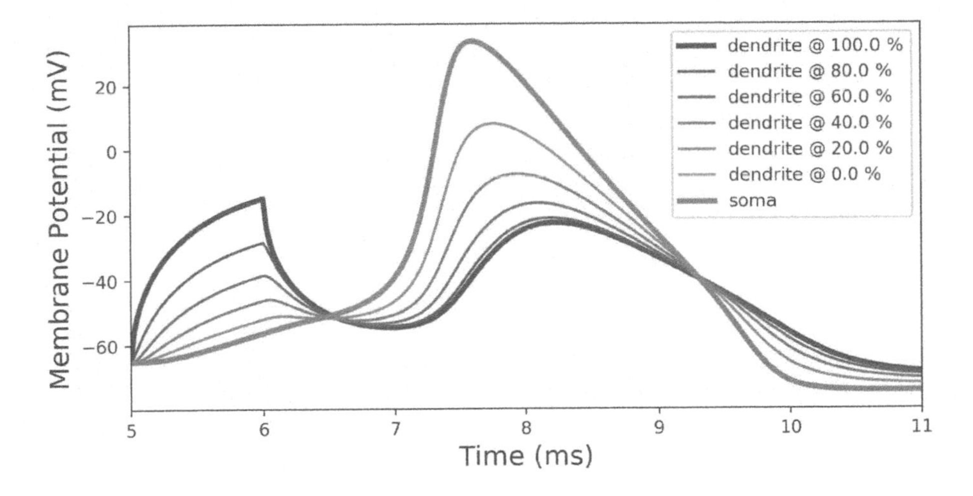

Figure 6.3 Simulating the cable equation with a supra-threshold stimulation. As we move further away from the stimulation point, the voltage level decreases. The point of current injection was labeled as 100%. The voltage level at the soma is sufficient to initiate a spike. Results show the spike's effect rippling back to the dendrite.

6.2.4 Partition length

As a guideline, cables should be discretized in intervals of 0.1 to 0.05 of the length-constant Λ which is defined with:

$$\Lambda = \sqrt{\frac{\frac{d}{4} \cdot R_m}{R_a}} \tag{6.15}$$

where d is the cable diameter, R_m is the membrane resistance per square meter and R_a is the axial resistance per meter [46]. This guideline accounts for the time it takes for a voltage to decay along a passive cable. A typical Λ value is 2 mm, leading to partitions of $\approx 100 \cdot 10^{-6}$ m in length, providing potential differences between partitions to differ by $\approx 5\%$. The greater the number of the partitions, the more accurate and complex the model can become. In **Figure 6.4**, the model described above was simulated with a different number of dendritic partitions. Results demonstrate that even with a simple simulation, the number of segments has a noticeable impact.

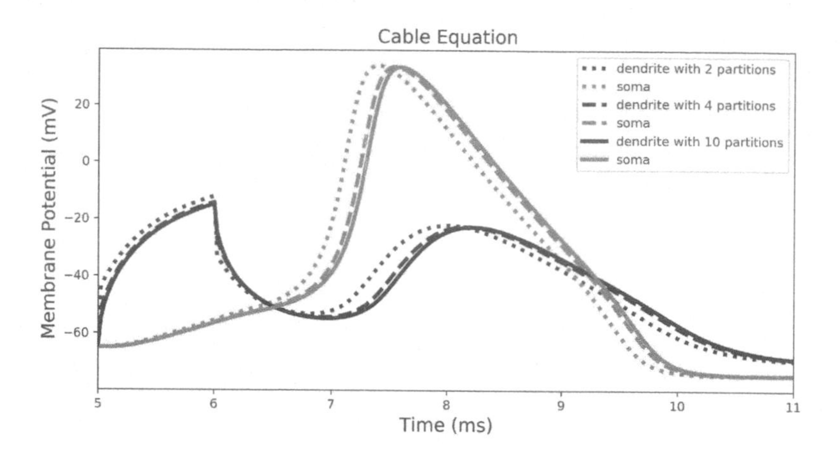

Figure 6.4 Simulating the cable equation with varying resolutions. Simulating the ball-and-stick neuron with varying number of partitions, providing slightly varied voltage traces.

6.3 THE COMPARTMENTAL MODEL

We can scale up the cable model by further connecting it in a branching architecture to other discretized cables, thus creating arbitrarily complicated morphologies (**Figure 6.5**).

As was demonstrated in **Figure 6.3**, injecting sufficient current to the dendrite will induce a response which will propagate forward, causing the soma to generate a spike. Connecting the soma with a synapse to another dendrite located on a different cell (neglecting the axon for convenience) will again induce a response which will propagate toward to the cell's soma, generating another spike. Repeating the process to a third "ball and stick" neuron will drive another spike in its soma. Simulation results are shown in **Figure 6.6**.

Note that while in Section 6.2, the guideline of using $100 \cdot 10^{-6}$ m partitions was given in heavily branching neurons, the short distances between the branches might require finer resolution, on the order of $10 \cdot 10^{-6}$ m or less.

6.3.1 Reconstructed morphology and dynamic

Designing simple neuronal structures with the compartmental model can be a straightforward task. However, biological neurons often have complicated shapes which might be intractable for manual structuring.

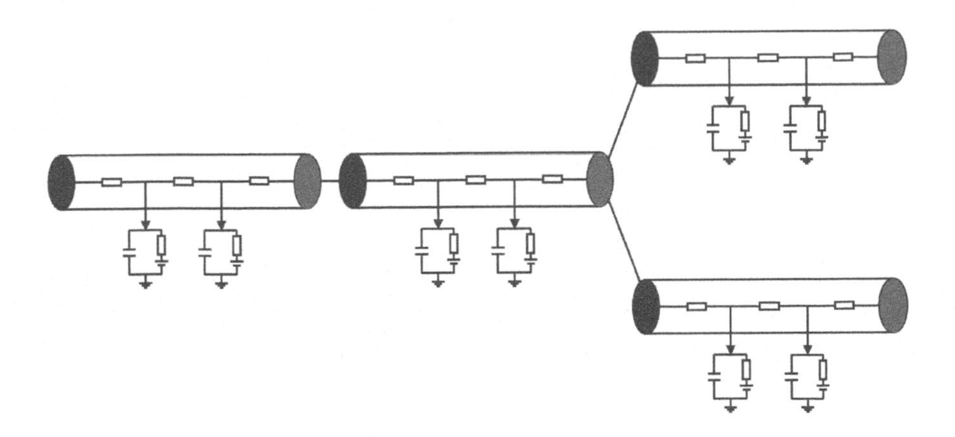

Figure 6.5 The compartmental model.

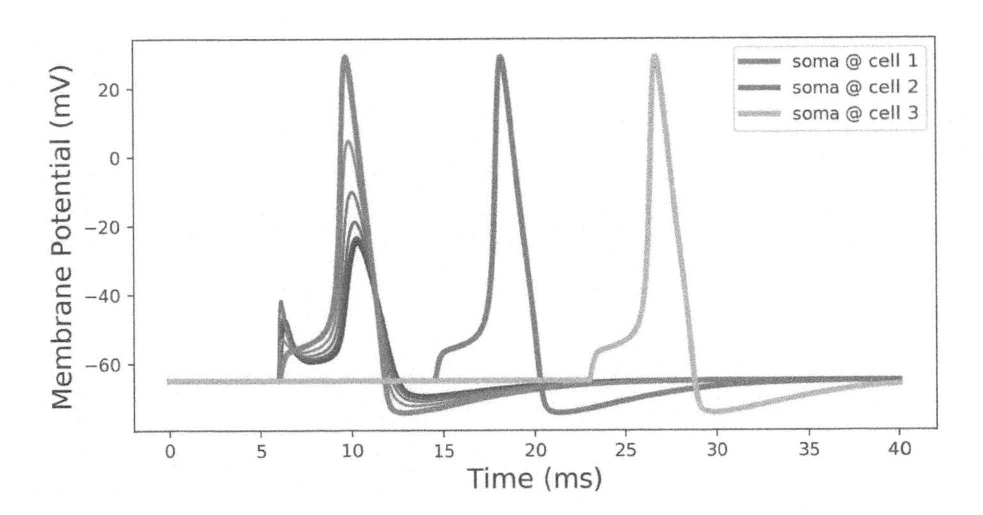

Figure 6.6 Simulating the compartmental model. Three ball-and-stick neurons were connected to each other (soma to dendrite). The first neuron in line was stimulated with a current injection. This caused a chain response in which all three neurons spiked in a delayed fashion.

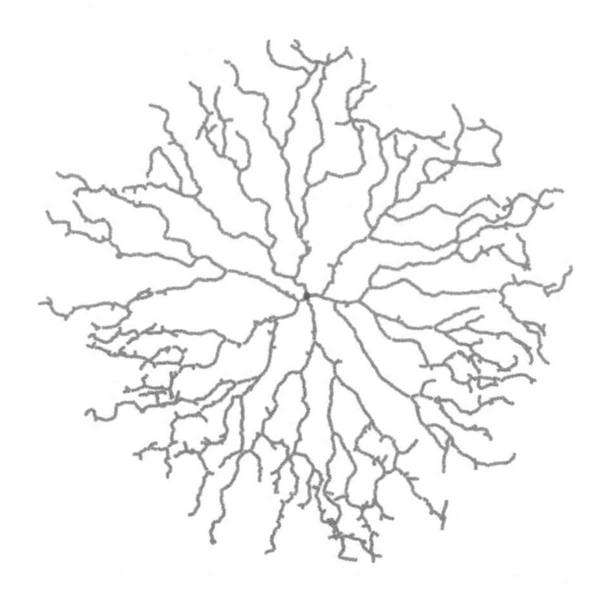

Figure 6.7 Reconstructed morphology of a SAC of a mouse. Derived from the open database: neuromorpho.org, model ID is NMO50993.

Fortunately, neuro-morphological details, gained with advanced imaging techniques (e.g., two photons and electron microscopy), can be used to reconstruct morphologies, allowing for accurate modeling of realistic neuronal shapes. Various reconstruction frameworks in which cellular morphologies can be autonomously or semi-autonomously derived from images are available [223]. A retinal SAC, which was reconstructed using the NEUROLUCIDA software [102] is shown in **Figure 6.7**. This geometry was retrieved from the open neural morphology database: neuromorpho.org. This SAC contains hundreds of millimeter-scale cables (or sections) and it can be imported to NEURON for simulation. Since this is a highly detailed morphological model, each section can be characterized differently, with different channel concentration and synapse distribution, to correlate with the biological, physiological details.

Importantly, since now we have a morphologically detailed model, we can track voltage response in each of the morphological sections, allowing a more comprehensive analysis of neuronal behavior (**Figure 6.8**).

6.4 CASE STUDY: DIRECTIONAL SELECTIVE SAC

SACs play a fundamental role in retinal directional processing. Particularly, as was mentioned in Section 5.3, SACs react differently to light

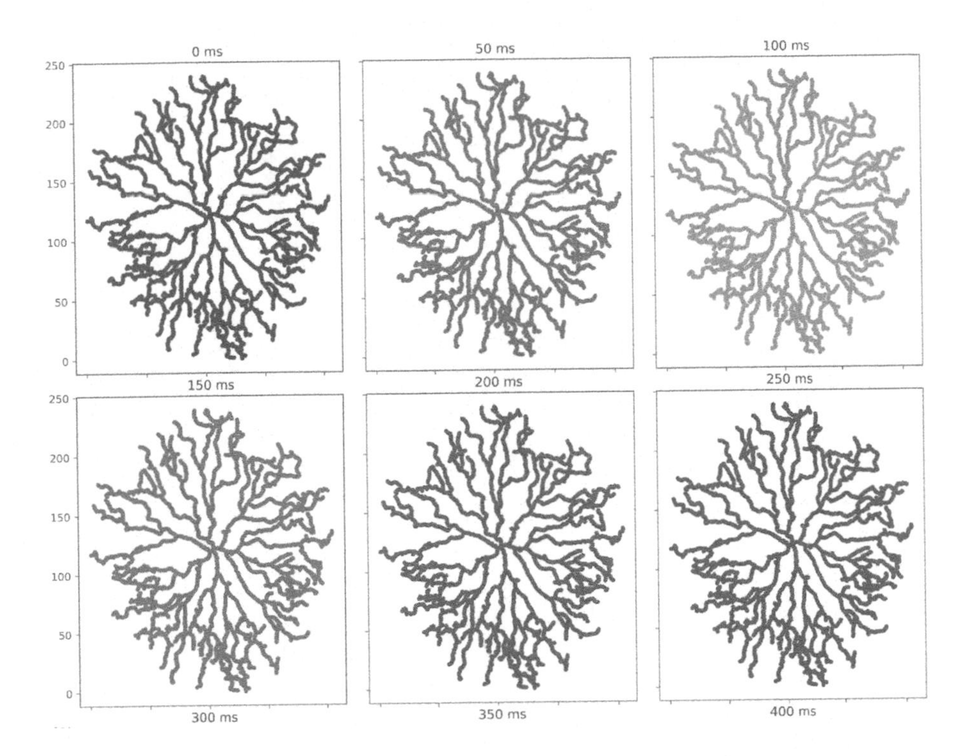

Figure 6.8 Morphologically detailed starburst amacrine cell response to an expanding ring of light, allowing voltage tracking in each of the morphological sections. See the detailed simulation description in Section 6.4.

coming from different directions, in a process tightly connected to their morphology and distributed biophysical characteristics. The SAC morphology, described in **Figure 6.7**, comprises hundreds of millimeter-scale cables (or sections) and it can be imported to NEURON for simulation. This is a highly detailed morphological model in which each section can be characterized differently, with different channel concentrations and synapse distribution, allowing us to better understand the mapping between its function and biological details. For example, we can design synapse density to follow a particular rule such as being inversely proportional to the distance from the soma; to define a high concentration of K^+ channels around the cell soma while setting the density of the N_a^+ and C_a^+ channels higher at the distal parts of the dendrites; and to establish a homogeneous resting potential throughout the cell (**Figure 6.9**) [292].

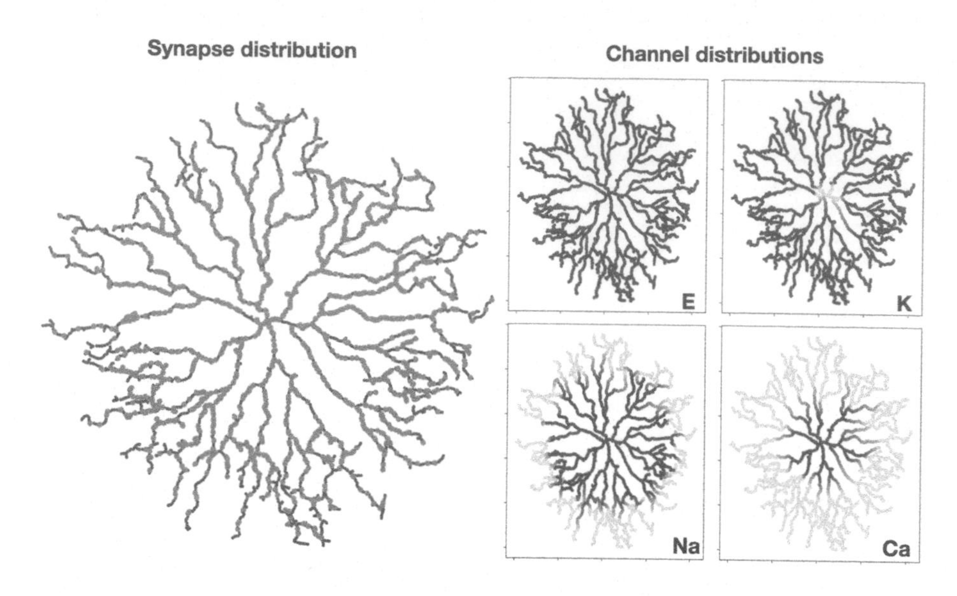

Figure 6.9 Reconstructed channel and synapse distributions. Synapse distribution is negatively correlated to their distance from the soma; K^+ is highly concentrated at the distal parts of the dendrites while N_a^+ and K^+ are concentrated at the distal parts; and resting potential is constant. Red dots indicate a synapse, and colors map channel concentration where yellow is high concentration and purple is baseline (arbitrary units).

SACs receive inputs from Bipolar Cellss (BCs). BCs are stimulated by photoreceptors – photo-sensitive cells which stand at the visual system's frontline. We can use NEURON to stimulate a SAC with expanding and collapsing rings of light (**Figure 6.10**) while introducing the light stimuli through BC-SAC synapse models. Designing a model for BC-SAC synaptic connection is not a straight-forward task. Physiological and anatomical data show a different response dynamic to light, with more sustained excitation in proximal processes and more transient excitation in distal processes [95, 157] (A reminder: in SAC, axons, and dendrites are synonymous and called processes). The precise Spatiotemporal synapse distribution contributes to the SAC's directional preference, as only during directional motion are the sequential activation of the sustained and transient inputs effectively integrated. The dynamic of BC-SAC synapses can be defined using a stochastic vesicle release mechanism [261] and regulated by a vesicle refiling rate, a vesicle release probability (Sections 4.1.4 and 5.4), and the dendritic distance after which the synaptic response changes from sustained to transient.

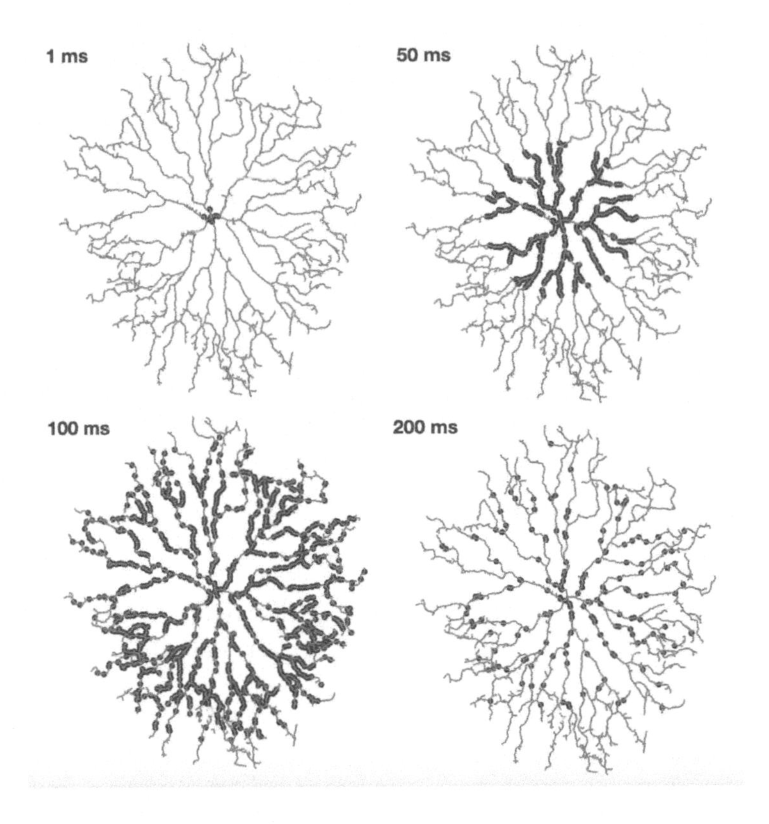

Figure 6.10 Stimulation of a reconstructed SAC with expanding ring of light, causing neurotransmitter release at the BC-SAC synapses.

In a recent work, we used a genetic algorithm to optimize the SAC's directional preference by changing the synapse distribution and dynamic. Notably, we showed that the distribution of excitatory synaptic inputs from BCs should be confined to the SACs proximal $\frac{2}{3}$ dendritic arbors, thus matching known anatomical constraints [19]. The simulated responses of individual SACs which were obtained using the genetic algorithm showed a high correspondence with experimental results.

Simulation results show that in comparison to a collapsing ring light, SAC voltage peak would be in higher amplitude in response to an extending ring of light (**Figure 6.11**). SAC's compute and communicate with analog signals, as they usually do not initiate spikes (some SACs do). In this simulation, we used a spike-like mechanism, or events, to model vesicle release.

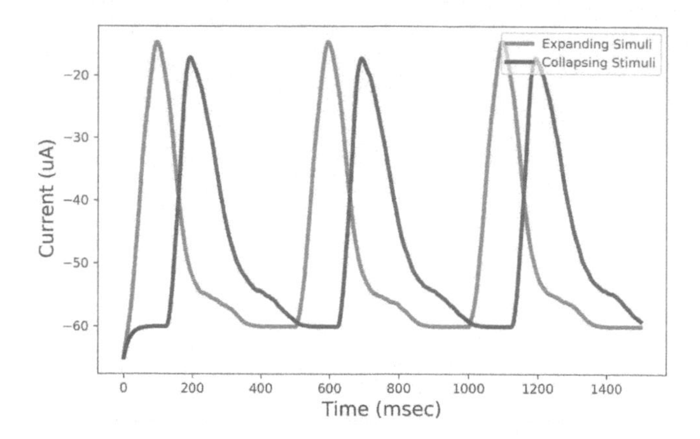

Figure 6.11 SAC response to expanding and collapsing rings of lights. In comparison to a collapsing ring light, an extending ring of light induces a stronger response.

This case study shows the importance of having highly detailed models of neurons and synapses to explain experimental findings and neuronal functions.

6.5 GLOSSARY

The cable equation: A mathematical formulation derived from an electrical model of the membrane and its intracellular and extracellular space, providing a quantitative description of current flow and voltage change both within and between neurons [127].

The compartmental model: Intricate morphological structures can be discretized into numerous interconnected compartments. Each compartment follows the cable equation.

6.6 FURTHER READING

- **Section 6.1**
  - The balance between detailed and abstraction in single-neuron models is described in [120].
  - Description of data structures for the description of detailed neuronal models is described in [284].
  - A description of NeuroML, a language for describing data-driven models of neurons and networks with a high degree

of biological detail is given in [104]. A detailed description of MorphML, the first level of the NeuroML standard, focusing on neuronal morphology data and model specification, is given in [65].

- Explore the Open Source Brain, at opensourcebrain.org. A database of morphologically detailed neuron models is at neuromorpho.org

- **Section 6.2**
  - The original work of William Thomson: "On the theory of the electric telegraph," in which he describes the cable equation for the first time is given in [275].
  - A detailed description of the core conductor theory and cable properties of neurons is given in [236].

- **Section 6.3**
  - In his paper: "Is realistic neuronal modeling realistic?," *Alon Korngreen* reviews the compartmental models and debates the level of details neuronal modeling should entail [10].
  - A seminal work by *Terrence Sejnowski* and colleagues in which compartmental models of reconstructed cortical neurons are used to reproduce a variety of firing patterns with neurons differ only in their dendritic geometry, is given in [188].

- **Section 6.4**
  - Throughout this case study and the next, retinal circuitry is discussed. Learn more about the retina by reading the following three important reviews (in order):
    * [195], a review by *Richard Masland* discussing the basic structure of the retina.
    * [300], a review by *Heinz Wassle* discussing the fundamental retinal circuitry.
    * [94], a review by *Greg Field* and colleagues discussing retinal information processing.

CHAPTER 7

Models of network dynamic and learning

Abstract

Neurons communicate with each other with vastly distributed synaptic connections. Each neuron might be connected to 1,000 other neurons, creating an amazingly complex dynamic. In Chapter 1.1, one of the most important paradigm shifts in neuroscience was described: the advancement from the neuron doctrine to neural networks. In neural networks, a function emerges from the joint activation patterns of the interconnected neurons. These neuron ensembles can generate emergent functional states which cannot be observed by studying the single entities they comprise. Importantly, as was described in Section 1.2.2, morphologically higher-order structures imposed by the morphological diversity within neuronal types impact emergent network activity. This chapter will demonstrate how detailed neural networks are described and simulated and how networks are utilizated for learning.

7.1 NEURAL CIRCUIT TAXONOMY FOR BEHAVIOR

The transition from the neuron doctrine to neural networks initiated a new paradigm in neuroscience. Neural networks became frameworks with high explanatory power with which high-level behaviors could be related to underlying neuronal mechanisms. In Section 1.1.3, the concept of networks and emergent behavior was succinctly discussed. Here, a few neural circuits are briefly discussed demonstrating underlying mechanical descriptions of high-level cognitive states.

In her article "Precision psychiatry: a neural circuit taxonomy for depression and anxiety," *Leanne Williams* describes a few key networks which underlie behavior [305]. Notably, in an attempt to guide clinical psychiatric practice, Williams draws a mechanical description of higher-level cognitive states, particularly those expressed in depression and anxiety.

Williams identified a few such circuits:

- **Default mode**. The network is described through the connections among the prefrontal cortex, cingulate cortex, and angular gyrus. This network is active when one is resting, when one's mind wandering, not focusing on anything concrete (rumination). It was shown that hyper-connectivity and over-activation of this network are related to abnormal rumination concerning depressive thoughts. Anatomical disruptions in the involved brain areas were associated with depressive disorders.

- **Salience**. The network is described through the connections among the cingulate cortex, insula, and amygdala. This network detects important environmental changes and signals the need for special attention. Abnormal connectivity within this network has been associated with depression, social anxiety, and panic disorder. This network is connected to the attention network (described below). Altered connections between the two might contribute to an avoidant personality in which one avoids over-stimulating situations.

- **Negative (punishment) and positive (reward) affects**. The "negative affect" network is described through the connections among the amygdala, brainstem regions, hippocampus, insula, and the prefrontal cortex. While certain parts of the network are responsible for emotional expressions, other parts of the network regulate negative emotions. Altered connectivity might affect one's ability to regulate threat and induce depression and anxiety. Hyper-connectivity to the default mode network might drive negatively-biased ruminations which can also induce depression. The "positive affect" network is described through the connections among the striatum and the prefrontal cortex. This network is responsible for reward processing and anticipation. Hypo-connectivity might lead to loss of sensitivity which might induce depression and anhedonia (lack of pleasure from social interactions, food, or sex). Hyper-connectivity might induce a context-insensitivity personality in which over-anticipation to reward results in reduced sensitivity to the surrounding context.

- **Attention**. The network is described through the connections among the frontal cortices, insula, inferior parietal lobule, and the precuneus. The attention network is responsible for alertness and sustained attention. It communicates with the default mode network, allowing alternating rumination and attention. Hypo-connectivity within this network is correlated with depression and social anxiety and can induce an inattentional personality.

These networks were defined using various experimental methods, such as functional Magnetic Resonance Imaging (fMRI). However, neuromorphic engineers are usually concerned with models with *deep* explanatory value, where neuronal models can be studied in a level of spikes and information processing.

7.2 RECONSTRUCTION AND SIMULATION OF NEURAL NETWORKS

7.2.1 Detailed modeling

In 2015, researchers in the Blue Brain project proposed a complete simulation of a $\approx 0.3\ mm^3$ cortical column in a mouse's neocortex [192]. This study simulated 31,000 neurons, categorized into 55 morphological and 207 electrical sub-types. The reconstructed neurons were strategically positioned and statistically connected in 3D space, according to biologically observed densities. Overall, this model had $8 \cdot 10^6$ connections and $37 \cdot 10^6$ synapses. In this biologically plausible model, neuron morphologies were reconstructed on which HH models were distributed encompassing 13 classes of ion channels. The Blue Brain project later reported the derivation of the full connection matrix (connectome) of the mouse neocortex, encompassing $88 \cdot 10^9$ synapses [241]. These models emphasize the scale detailed simulations target.

Such detailed biological models of neural networks are guided by bottom-up modeling. They are driven by extensive biological descriptions and an immense amount of experimental data. Unfortunately, data and biological descriptions are often incomplete or inaccurate. Therefore, numerous optimizers were developed to determine the right set of parameters necessary for a model to reproduce experimentally observed firing patterns [68]. In Section 6.4, a morphologically detailed simulation of a SAC with a biologically-demonstrated directional preference was described. We derived a set of relevant parameters using a genetic algorithm, such that the cell would exhibit experimentally observed behavior. Later, in Section 7.3, we will show that while SAC's directionality is crucial to a directional response of Direction Selective Ganglion Cells' (DSGCs) which they regulate, SACs' participation in an inhibitory network drives these DSGCs to be even more directional.

How vital are detailed neuronal morphologies to our understanding of neuronal function? To what extent do fine morphological details of neurons affect their spiking activity? Various studies, including the one conducted at the *Idan Segev* lab of the Blue Brain project by *Oren Amsalem* and colleagues, aimed at simplifying a cell's morphological complexity while preserving spiking characteristics [13]. A reduced cell's morphology preserves the original reconstructed cell's main morphological and branching characteristics while accelerating simulation 40 to 250 folds. Finding the balance between morphological plausibility and computational demand is key to efficiently simulate large-scale, biological models.

7.2.2 Simulating detailed models

Simulating the network specified in Section 7.2 takes $\approx$ 15 minutes to resolve on standard hardware (2.6GHz six cores Intel i7 processor and 16GB RAM). Scaling this model up to 10^6 neurons, assuming linear time scaling, would lead to $\approx$ 2 years-long simulation, demonstrating the intractability of such large-scale models. Large-scale modeling often requires an immense amount of computational resources (see also Section 1.2.2). Thus, a key feature in simulation design is its scaling capability. One solution might emphasize distributed computing. The Blue Brain's simulations, for example, are executed with an extended version of NEURON which supports multi-core processing. In various conditions, even a single neuron can be allocated for evaluation on different processors. It was shown that a reconstructed morphology could be distributed to more than ten processors, with a small overhead [124]. Since GPU can offer thousands of simple computing cores, they can accelerate large scale networks. Leveraging GPU computing for large-scale neural networks is gaining increased traction as some studies demonstrated its comparable performance to super-computers and neuromorphic hardware [161]. Accelerating large-scale neuronal simulations is also critically dependent on software optimization [63]. Therefore, an important stepping stone in this field is to co-design the hardware and software, allowing integrated optimization. These algorithms' details are outside of this book's scope but are referenced in this chapter's further reading.

In the context of high-performance computing, there are two common notions of scalability:

- **Strong scaling** in which the time to a solution of a fixed problem is measured as a function of the number of processors used.
- **Weak scaling** in which the average processing load of each processor is measured as a function of the number of processors used.

While tremendous effort and engineering ingenuity were utilized to exhibit strong and weak scaling in super-computers, scaling is almost a given in neuromorphic hardware. From the scientist's perspective, one of the field's key drivers is the potential of using neuromorphic hardware to simulate large-scale biological plausible neural networks, ideally in real-time and with low energy consumption. While neuromorphic systems, such as IBM's TrueNorth, were not designed to execute detailed biological plausible neural models, it was shown that models of neuron dynamics (such as the HH model) could be mapped and deployed on the chip [181]. This was accomplished by rate coding numerical values with spikes and performing mathematical operations by configuring the TrueNorth's neurons with specific sets of parameters.

Numerous reasons were raised to justify the tremendous expenses and efforts dedicated to large-scale simulations of biologically plausible models, spanning both practical and fundamental implications. It was suggested that realistic simulations would help reveal the functional significance of the intricate organization of neurons and synapses; allow a better understanding of the ways information is propagated and processes in the brain; provide a window to neuro-developmental processes and brain disorders; and provide a framework for drug discovery. Most importantly, some will claim that this will pave the way to answer the eternal question of what makes us conscious beings.

Some of the simulations provided insights into brain function, providing a spectrum of hypotheses regarding the control of network states (e.g., shifting from synchronous to asynchronous activity). However, aside from exhibiting and glorifying impressive technological advancements in terms of efficient distribution of models on many thousands of computing cores, these immense simulations have yet to be proven to have functional emergent dynamics or pose new information processing capacity [86]. As most simulations conclude with the observation of non-randomized patterns of activity, we are only starting to uncover the tip of this very deep iceberg.

7.3 CASE STUDY: SACS' LATERAL INHIBITION IN DIRECTION SELECTIVITY

In the case study described in Section 6.4, the directional response of a SAC was described. It is further known that the SAC network has an essential role in the DSGCs directional response which it regulates. Indeed, a dense converge of SACs was shown to enhance the directional response of DSGCs [214].

To reconstruct a network, we can define a template of a particular morphology, instantiating it multiple times, creating populations of identical neurons. For example, a dense coverage of SACs can be constructed by placing identical SACs in overlaying grids. Soma plot (only cell somata are shown) of such 2D grid is shown in **Figure 7.1**.

By iteratively instantiating SAC morphology at the appropriate locations defined above, a network topology of SAC plexus can be defined (**Figure 7.2**).

In the SAC plexus, intersecting SACs inhibit each other with GABA-based (GABAergic) synaptic connections. These synapses can be defined at the intersections of neighboring cells. Once network topology is defined, synapse location is calculated. SACs are driven by BCs. BCs transverse illumination from photoreceptors to DSGCs and their regulatory neurons, SACs. The reconstructed network with both inhibitory SAC-SAC synapses (colored green and yellow, termed lateral inhibition) and excitatory BC-SAC synapses (colored red) is shown in **Figure 7.3**.

Similar to what was demonstrated with a single SAC (**Figure 6.11**), we can use the SAC plexus to study the directional response of the network by stimulating it with expanding and collapsing rings of light (**Figure 7.4**).

Simulation results show that the SACs lateral inhibition increased the directional response, both in terms of amplitude difference and activation dynamic (rise and fall times) (**Figure 7.5**).

DSGCs are regulated by a SAC plexus. This regulation scheme drives DSGCs' directional response, thus establishing retinal directionality. To explore these phenomena, we connected the SACs to a ganglion cell and monitored their response to directional stimuli (bars drifting in two opposite directions). First, the DSGC was connected to the SAC plexus

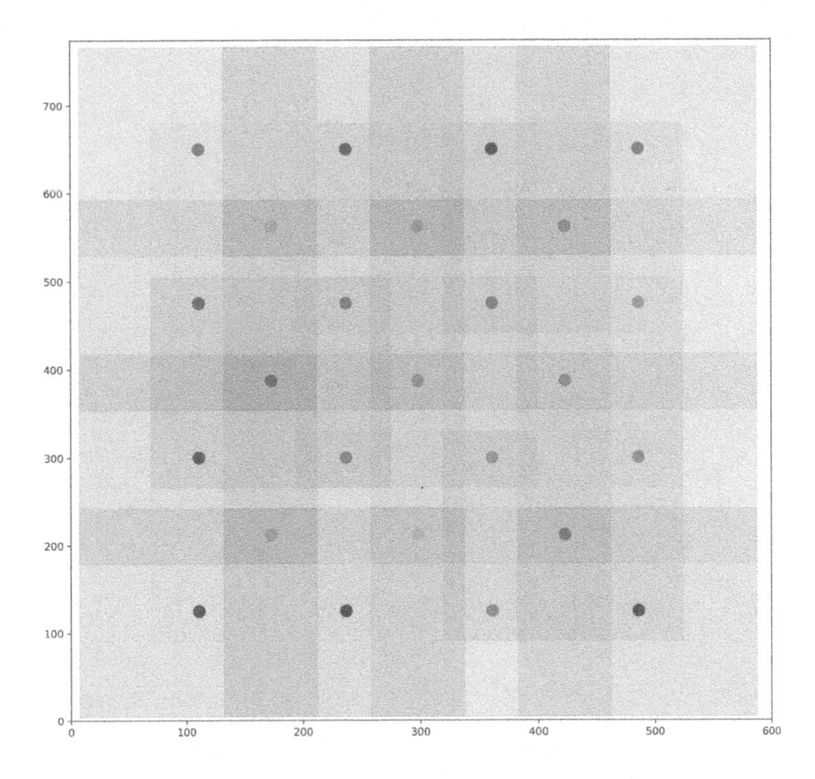

Figure 7.1 Somata plot of overlying 2 grids of cells. Cell size was derived from the SAC morphology shown in **Figure 6.7**. Numbers are in 10^{-6} m.

via randomly distributed GABAergic (inhibitory) synaptic connections. With this random connectivity scheme, the DSGC did not produce a directional response (**Figure 7.6, A**). Next, we determined SAC-DSGC inhibitory synapses according to a pre-specified connectivity rule. In particular, it has been shown that the connectivity scheme of SAC processes onto a target DSGC is based on the orientation of the individual SAC process relative to the DSGC Preferred Direction (PD). SAC processes preferentially connect to DSGCs with a PD anti-parallel to the process. To implement this rule, we set the probability function of SAC and DSGC synapse formation to follow the inverse cosine similarity between the vectorized direction of the SAC's process (relative to the soma) and the PD of the DSGC (**Figure 7.6, B**). When running the simulation, even in the absence of lateral inhibition within the SAC network, PD motion evoked stronger depolarization in the simulated DSGC than null direction motion (**Figure 7.6, C**). However, the DSGC's directional

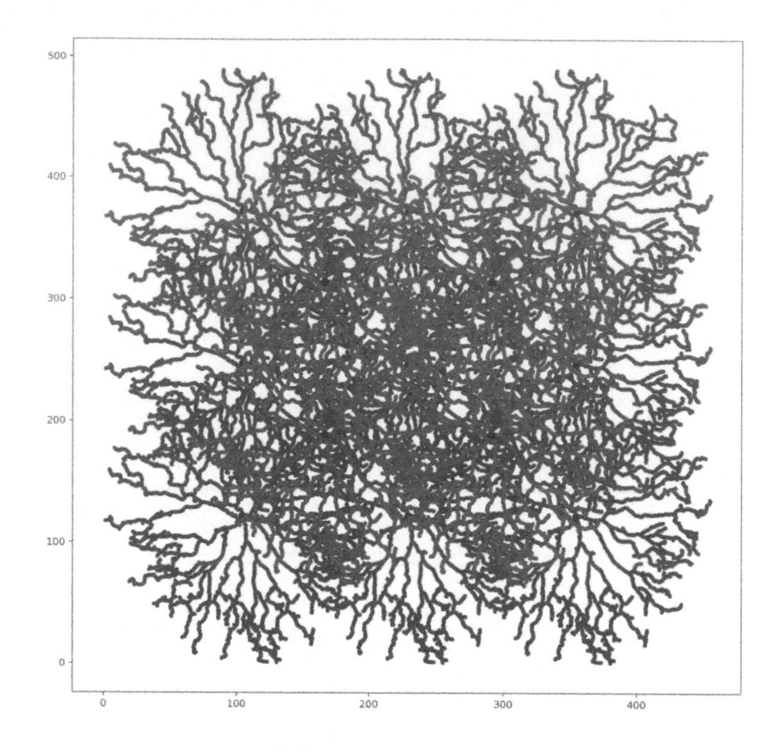

Figure 7.2 Morphologically detailed dense plexus of starburst amacrine cells. Cells were defined using the cell template shown in **Figure 6.7**.

preference was further enhanced upon inclusion of lateral inhibition in the SAC network (**Figure 7.6, D**). Simulation results show that the network's lateral inhibition increased its directionality (**Figure 7.5**).

Simulation results are summarized in **Figure 7.7**. Directionality D index was assessed by:

$$D = A_{PD} - A_{ND} \tag{7.1}$$

where A_{PD} is the area under the voltage trace at the PD, and A_{ND} is the area under the voltage trace at the null direction. Areas were calculated above 10 mV threshold voltage.

The network encompasses 13 morphologically detailed cells, with $\approx$ 10,000 sections, $\approx$ 28,000 segments, and $\approx$ 20,000 synapses. This is, however, a reasonably small-scale simulation. A larger-scale simulation would require a specialized super-computing hardware.

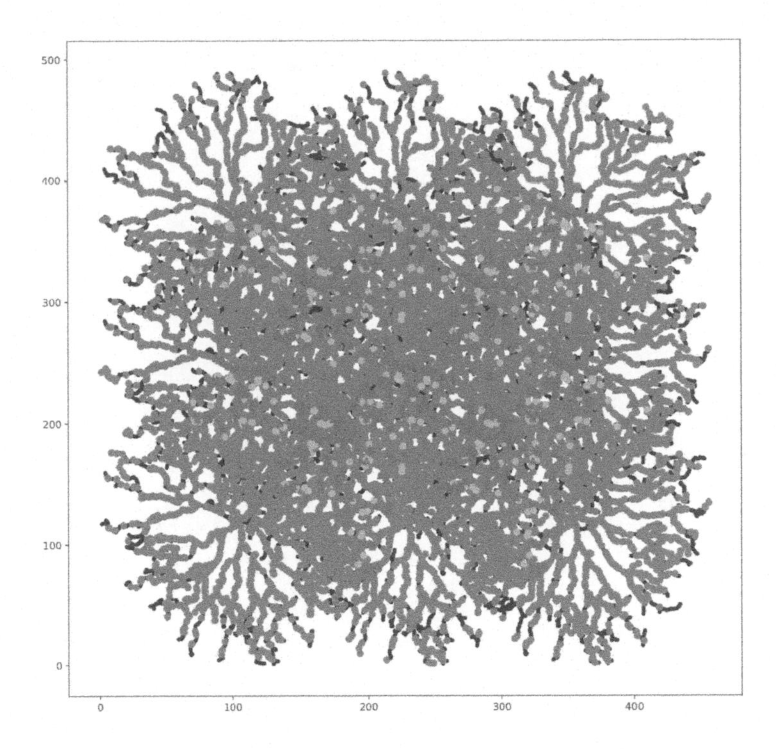

Figure 7.3 Synapse plot in a SAC network. Inhibitory GABAergic SAC-SAC synapses are colored in green and yellow (source and target respectively; randomized selection) and excitatory BC-SAC synapses are colored in red.

7.4 NEUROMORPHIC AND BIOLOGICAL LEARNING

One of the essential characteristics of life is the organism's ability to adapt to a changing environment dynamically. The most important strategy for adaptive behavior is learning. Learning in the brain takes place through careful modulation of synaptic connections (see Section 3.1). Generation and degradation of synapses and changes in dendritic and axonal arbors also support learning. Notably, the brain can even dynamically allocate more of its capacity to certain tasks according to task prevalence. The brain is, therefore, highly plastic, as it physically and physiologically changes in response to experiences.

While ANNs are typically organized in interconnected layers, the brain's neural networks are organized differently, as there is usually no direct mapping between one "layer" of neurons to another. Even in the seemly layered structure of the cortex (Section 1.2.2), there is no

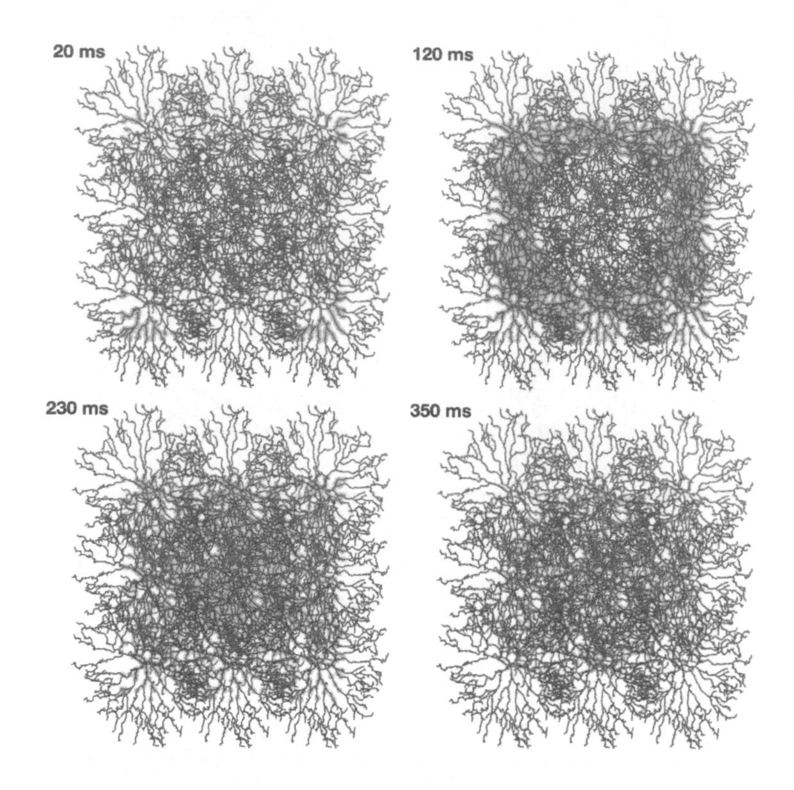

Figure 7.4 Stimulation of a reconstructed network of SACs with an expanding ring of light, causing neurotransmitter release at the illuminated BC-SAC synapses.

precise "forward" mapping between layers. Furthermore, in the cerebral column, different cell types characterize each cortical layer, and cortical areas form a diverse set of connections to both cortical and sub-cortical regions [192]. However, elucidating possible learning strategies taken by the brain might drive the development of learning algorithms for SNNs (from the perspective of the algorithm developer) or give some insights on biological learning (from the perspective of the scientist).

Here, the discussion will be divided into:

- **Supervised learning**, in which error signals are used to modulate the neural network towards the optimization of a desired behavior.
- **Unsupervised learning**, in which a behavior is defined without explicitly defining it, via pattern recognition.

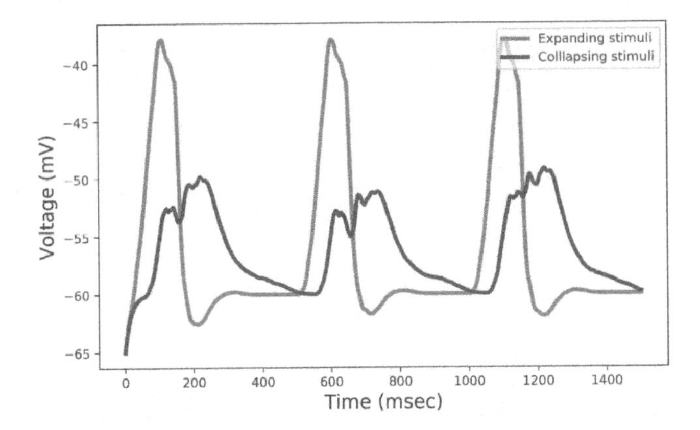

Figure 7.5 Network of SAC response to expanding and collapsing rings of lights. In comparison to a collapsing ring light, results show distinct amplitude and dynamic. These effects are more pronounced here than with a single cell, as was shown in **Figure 6.11**.

7.4.1 Biological backpropagation-inspired learning

With ANNs, synaptic weights are modulated such that some loss function is minimized, most commonly via the *backpropagation* learning algorithm. Back-propagation learning is the backbone of modern machine learning. It is underlying ANNs' productivity and it is used for diversed aims - from computer vision to language modeling. With backpropagation, an error signal propagates back from the network's outputs to its inputs while adjusting connections' weights. In this *supervised learning* methodology, both error and desired behavior are provided, allowing network training such that it would ideally provide a generalized model that maps inputs to a behavior. While the relevance of backpropagation to biological learning is still widely debated, a growing body of works shows that the brain can use a similar learning mechanism. *Timothy Lillicrap* outlines some difficulties that arise from the attempts to apply traditional backpropagation in biological learning [175]:

1. Backpropagation relies on forward and backward passings through the network. In the forward passing, the network's behavior is monitored as input data and is introduced through its nodes (neurons). In the backward passing, the network's synaptic weights are modulated, such that the forward pass would be closer to the desired behavior. Therefore, to conclude a new and better weight scheme

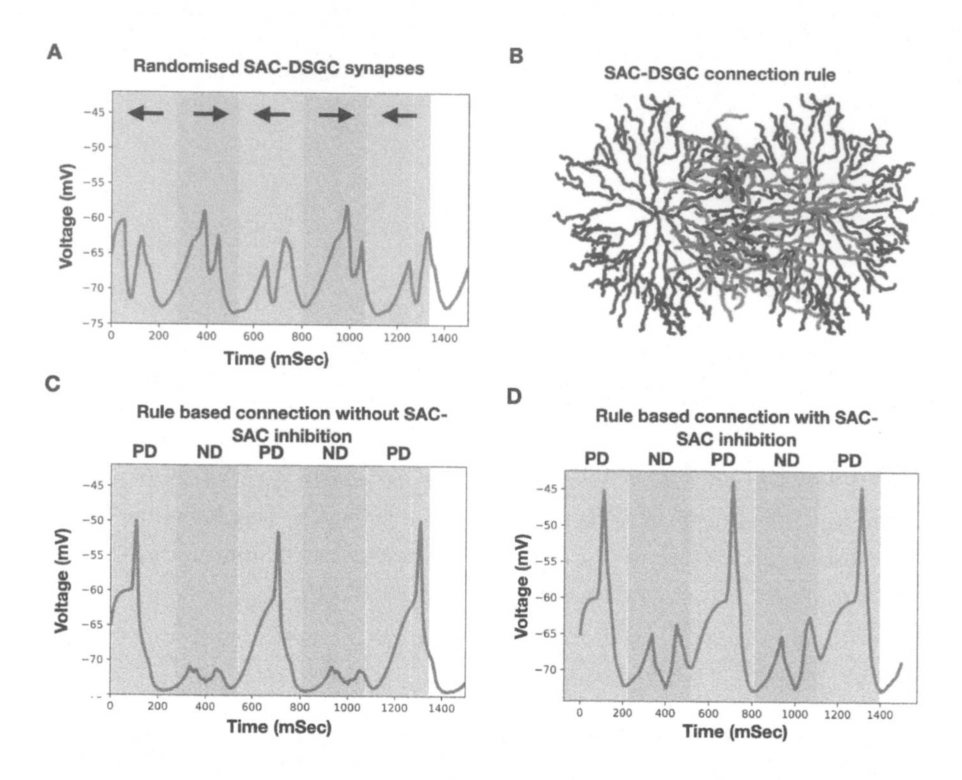

Figure 7.6 DSGC was connected to the SAC plexus and exposed to a drifting bar alternating between the Null (ND) and PD. Voltage traces are the DSGC's somatic voltage; A. When the DSGC was randomly connected to the SAC plexus, SAC lateral inhibition was set to 0.05 nS; direction selectivity was not observed; B. Schematic of two neighboring SACs (morphology is colored with purple; synapses between the SACs are not shown for simplicity's sake) are connected with GABAergic synapses (red) to a DSGC (blue). SAC-DSGC synapses were defined with higher probability when the SAC's process and the DSGC's PD are anti-parallel; C. When the connectivity was asymmetric and SAC lateral inhibition was set to 0 nS, DSGC's response was directional; D. Direction selectivity is sharpened when SAC lateral inhibition is enabled.

for the network, its forward and backward paths must be symmetric. Having two different paths with the same weight scheme is not biologically plausible. This "weight transport problem" is thought to be one of the main reasons to question the biological plausibility of backpropagation [172].

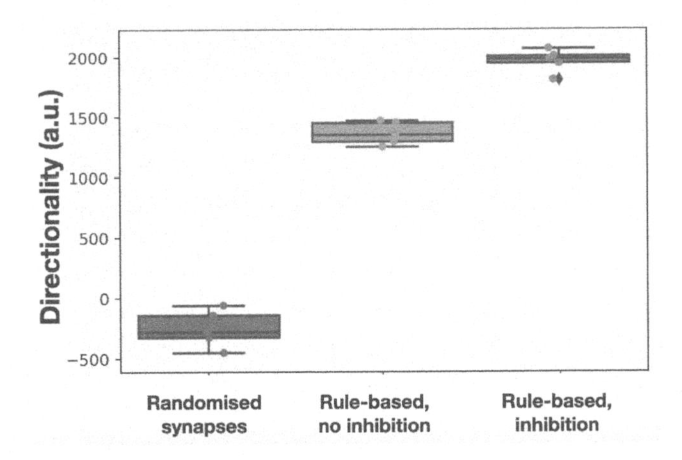

Figure 7.7 Compared directionality in different retinal models: with randomized SAC-DSGC synapses (blue), rule-based SAC-DSGC synaptic distribution (orange) and rule-based SAC-DSGC synaptic distribution with lateral SAC-SAC inhibition (green).

2. With ANNs, the propagating error signals are signed (can be positive or negative) and are prone to vanish (to very small values) or to explode (to very big values). Conveying positive or negative signals through a BNN is problematic. Proposals, such as setting a certain spike rate θ as 0, where a higher rate ($f > \theta$) indicates a positive error signal and a slower rate ($f < \theta$) indicates a negative signal, were deemed inadequate as integrating several such inputs in a synapse that is hard to compute and requires implausible memory. Moreover, ANNs violate Dale's law which claims that a biological neuron can be either excitatory or inhibitory (and not both) (See Section 4.1.4).
3. In backpropagation, error signals are used to modulate synaptic weights while not influencing the activity states of neurons. In biology, feedback can alter the neurons' response dynamic as it alters their weight.

For a biologically plausible backpropagation, some neurons have to represent neural activity differences to encode errors and to drive synaptic changes, a principle termed Neural Gradient Representation by Activity Differences (NGRAD) [175]. The idea of utilizing temporal differences between neuronal activities to drive synaptic updates has been explored with different mechanisms, including:

- Boltzmann machines [4]. A Boltzmann machine is a network of hidden and visible binary units which are symmetrically connected via undirectional weighted edges. Unlike classic gradient descent, learning with Boltzmann machines is biologically plausible due to the "locality" of its governing learning rule. It is local because weight updates between every two nodes do not require information other than the nodes themselves. From the perspective of the algorithm designer, Boltzmann machines have not been proved useful as time to solution grows exponentially with the machine's size (thus driving the development of restricted Boltzmann machines).
- Dendritic computing [252]. In this learning mechanism, commonly inspected on models of cortical pyramidal neurons, weight updates are driven by local dendritic computation. Pyramidal neurons have two types of dendrites: apical and basal. While apical dendrites are distal in reference to the cell's soma (and reside in layer 1 of the cortical column), basal dendrites are closer to the soma (thus receiving input from neighboring cells). In this model, both long-range excitatory feedback from apical dendrites and local inhibitory predictions from basal dendrites are integrated to produce an error signal. This error signal can be used to modulate synaptic weight at the synapses of the basal dendrites.
- Local Hebbian synaptic plasticity [302]. In this model, termed a predictive coding framework, a local version of Hebbian learning (which to be discussed in Section 7.4.2.1) can be used to implement backpropagation-like learning.

NGRAD are characterized by their ability to induce weight updates in a backpropagation-like manner using locally available signals. Thus learning is driven without having explicit error propagating through neuronal layers. Supervised biological learning is responsible for a vast array of human capabilities ranging from playing in a gaming console to playing guitar.

The brain has a "prior knowledge" regarding its immediate environment, encoded with innate mechanisms (brain regions, connectivity, and initial synaptic weights), cultivated by evolution. It gives the individual a "head start" in his race for survival, enabling him to rapidly adapt to his environment with a minimal number of examples for desired behaviors.

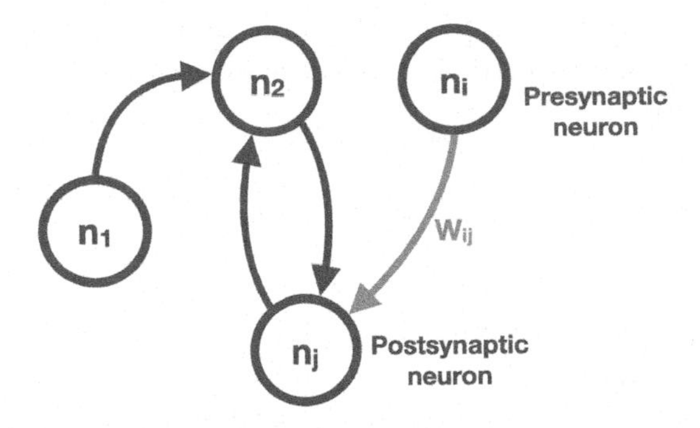

Figure 7.8 With Hebbian learning, synaptic weights are locally adjusted, as the information required for update is embedded within the activity of the pre- and postsynaptic neurons, designated here with i and j, respectively.

7.4.2 Biological unsupervised learning

7.4.2.1 Hebbian learning

Most biological learning takes place through *unsupervised learning*. One of the classic reported mechanisms for unsupervised biological learning is *Hebbian Learning*, proposed by *Donald Hebb* back in 1949 [113]:
"When an axon of cell A is near enough to excite cell B and repeatedly or persistently takes part in firing it, some growth process or metabolic change takes place in one or both cells such that A's efficiency, as one of the cells firing B, is increased."
This learning rule is commonly succinctly phrased as "cells that fire together wire together."

Mathematically, Hebbian learning can be formulated as:

$$\frac{\delta w_{ij}}{\delta t} = f(w_{ij}, a_i, a_j) \tag{7.2}$$

where $\frac{\delta w_{ij}}{\delta t}$ is the magnitude of change for the modulation of connection weight w_{ij}, i is a presynaptic neuron, j is a postsynaptic neuron, and a_x is the spiking rate (or activity) of neuron x (**Figure 7.8**).

Hebbian learning bring forth a requirement for joint activity of neurons i and j for a synaptic weight change to occur, termed c_{ij}^{corr}, giving the canonical formulation:

$$\frac{\delta w_{ij}}{\delta t} = c_{ij}^{corr} \cdot a_i \cdot a_j \tag{7.3}$$

When c_{ij}^{corr} is negative, the resulted learning behaviour is "anti-Hebbian," as a joint activity of two neurons would cause their synaptic connection to weaken. When c_{ij}^{corr} is positive, synapses can only be strengthened and all synaptic weights will increase to infinity, resulting with an instability. However, c_{ij}^{corr} can be defined such that synaptic weights would eventually saturate at some upper maximum value w_m. One way to achieve such saturated behaviour is to define c_{ij}^{corr} as a function of w_{ij} by defining $c_{ij}^{corr} = \lambda$ when $0 < w_{ij} < w_m$ and $c_{ij}^{corr} = 0$ otherwise. λ is defined as a learning rate. Another approach would be to define a "soft" transition, by defining:

$$c_{ij}^{corr}(w_{ij}) = \lambda(w_m - w_{ij})^{\beta} \tag{7.4}$$

As β approaches 0, this model behaves more similarly to the "hard" transition approach.

An important variant of Hebbian learning dictates the depression of a synaptic connection, when joint activation is zero (e.g. no stimulation) by defining:

$$c_{ij}^{corr}(w_{ij}) = \lambda(1 - w_{ij})a_i a_j - \lambda_d w_{ij} \tag{7.5}$$

Hebbian learning was demonstrated computationally by *Kenneth Miller* back in 1989, showing the emergence of ocular dominance columns in the 4th layer in the visual cortex [210]. At birth, spontaneous activity occurs in each retina, where neighboring neurons - located in the same eye - tend to be wired to neighboring neurons in the cortex. Simulating Hebbian learning-driven ocular dominance reveals biological-like "patches," demonstrating the development of dominance columns over time. This simple example demonstrates the importance of having a model for understanding a biological phenomenon.

7.4.2.2 *Spike timing-dependent plasticity*

One important spike-tailored Hebbian-based learning rule is STDP. STDP was demonstrated to be important in a wide spectrum of biological neural networks and it revolves around the importance of the

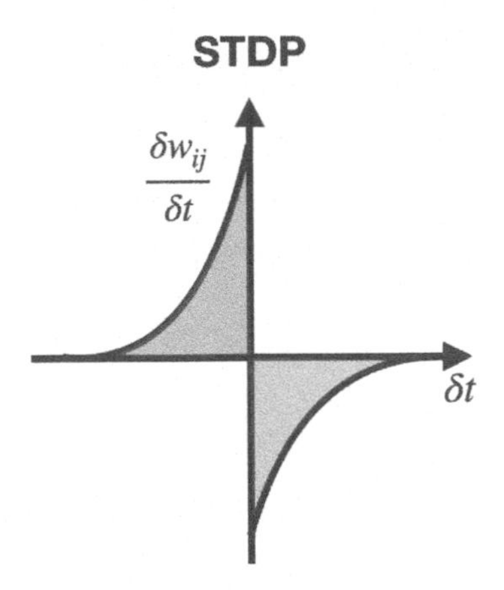

Figure 7.9 With STDP, a positive increase in a synaptic weight occurs when the presynaptic spike precedes the postsynaptic spike (LTP, colored red), and vice versa (LTD, colored blue).

temporal order of pre- and postsynaptic spiking for synaptic modification [49].

STDP defines for a pair of pre-synaptic and post-synaptic cells a critical $\approx$ 20 msec-long time window in which spiking activity can modulate their connection weight. Formally, given a presynaptic spike at time $t = t_{pre}$ and a postsynaptic spike at $t = t_{post}$, and $\delta t = t_{pre} - t_{post}$, we define STDP using:

$$\delta w^{+} = A^{+}(w) \cdot e^{\frac{\delta t}{\tau^{+}}} \tag{7.6}$$

When t_{pre} precedes t_{post}, such that $t_{pre} < t_{post}$ and δt is negative, the synaptic connection between the neurons will strengthened. Alternatively, when $t_{post} < t_{post}$:

$$\delta w^{-} = -A^{-}(w) \cdot e^{-\frac{\delta t}{\tau^{-}}}. \tag{7.7}$$

For both equations, A^{+} and A^{-} are functions of the synaptic weight. This model is limited to perform a weight update on pairs of spikes. More complicated models define a window of time in which pre- and postsynaptic spikes can be considered related. STDP rule is visualized in **Figure 7.9**.

To showcase STDP, we will create a two-neurons model in which spiking behavior is defined to follow a Poisson-guided distribution and a

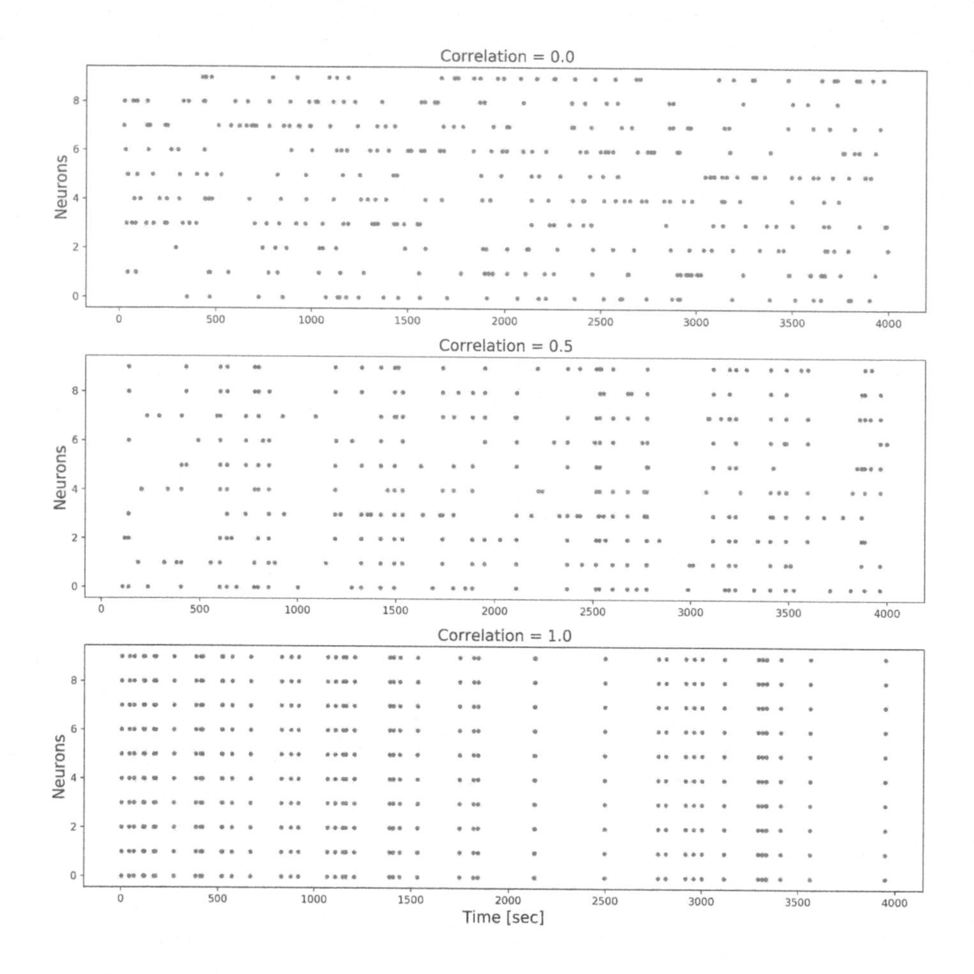

Figure 7.10 Three spiking patterns with different correlation values (1, 0.5, and 0), generated by 10 neurons.

correlation factor. Here, three groups of 10 neurons with different correlation values are shown in **Figure 7.10**. As the correlation between the neurons' spikes increases, the closer and more related they are to each other. Implementation details are given on the book website (spiking behavior was generated using the Brian simulator [106]).

Monitoring the modulation of synaptic weight between each two neurons within each spike group following STDP was examined. Results are shown in **Figure 7.11**

STDP is a biologically plausible learning rule which is routinely used for real-time learning in neuromorphic systems (see Chapter 13) [226].

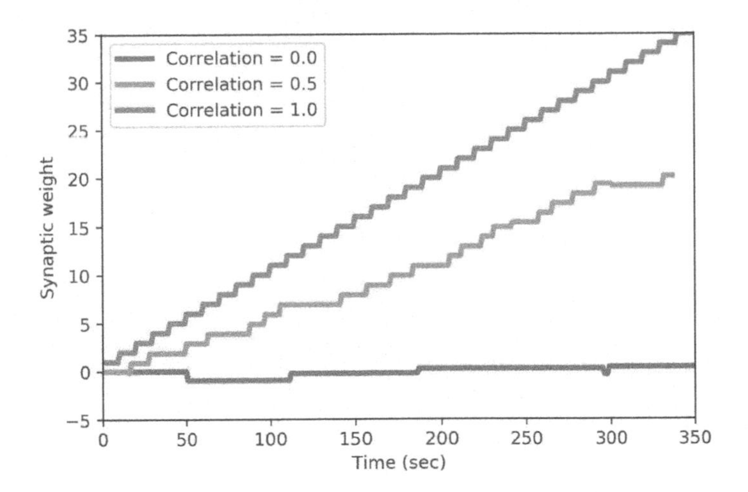

Figure 7.11 Synaptic weight between two neurons within three different groups of neurons, each group with a different level of correlated behavior. In this model, $\tau^- = \tau^+ = 0.01$ and $A^-(w) = A^+(w) = 1$.

STDP was implemented on the SpiNNaker [75], the Loihi [176] and the TrueNorth [77] architectures. Neuromorphic systems can provide a means with which learning strategies with biologically plausible SNNs can be studied on a large-scale, thus providing immense value to our knowledge of biological learning [76]. For example, *Saeed Reza Kheradpisheh* demonstrated the use of STDP-driven spiking neural networks to quickly recognize objects in natural images [156]. Such experiments may prove important for developing a complete theory regarding biological vision, explaining its processing speed and energy efficiency.

A detailed, formal presentation of Hebbian learning and its variants is given in [100]. Two variants, the Bienenstock-Cooper-Munroe (BCM) and Oja's learning rules, are briefly described below.

7.4.2.3 BCM learning

BCM is an important variant of Hebbian learning [36]. The BCM learning rule follows STDP, where the pre- and postsynaptic neurons are weakly correlated, and only nearest-neighbor spike interactions are taken into account [143]. The BCM rule stabilizes Hebbian learning by reducing synaptic weights when the postsynaptic neuron becomes too active

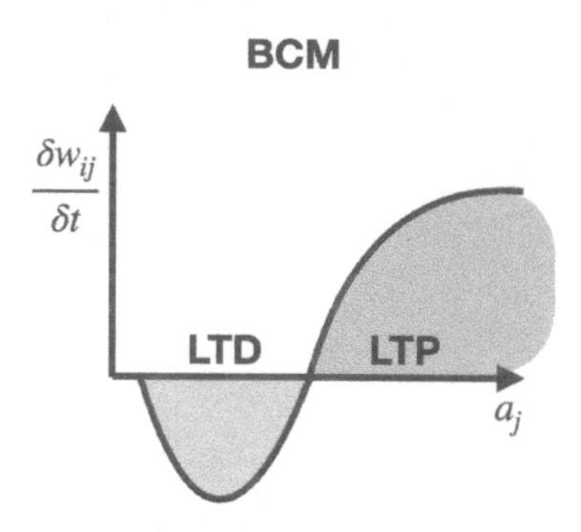

Figure 7.12 With the BCM, synaptic modification is characterized by two thresholds separating non modifying, positive and negative activity levels. LTP is colored in red and LTD is colored in blue.

and it is defined using:

$$\frac{\delta w_{ij}}{\delta t} = \psi(a_j - a_\tau)a_i \tag{7.8}$$

where ψ is a non-linear function and a_τ is a function of the average postsynaptic spike rate a_j. BCM is described with a_τ, a threshold activity, under which synapse is undergoing weakening (or depression), achieving Long Term Depression (LTD). When a_j is higher then a_τ, the synapse undergoes strengthening (or potentiating), achieving Long Term Potentiation (LTP) (**Figure 7.12**). Sometimes, another threshold a_0 is used to denote postsynaptic activity under which the synapse weight remains the same.

7.4.2.4 *Oja's learning*

Another way to stabilize the Hebbian rule is through *weight normalization*, constituting the Oja's rule [221] which is defined using:

$$\frac{\delta w_{ij}}{\delta t} = \lambda(a_i a_j - w_{ij} a_j^2) \tag{7.9}$$

Here, weights are normalized by: $\sum_i w_{ij}^2 = 1$. Therefore, when some synaptic weights increase, others decrease.

7.5 GLOSSARY

Back-propagation: A prominent algorithm for training ANNs in which a network's weights are modulated following a gradient of an error function.

BCM learning: Hebbian learning-based rule according to which a neuron will undergo LTP if it is in a high-activity state or LTD if it is in a lower-activity state.

Hebbian learning: Activity-dependent synaptic plasticity where correlated activation of pre- and postsynaptic neurons leads to the strengthening of the connection between the two neurons [56].

Long Term Depression: Activity-dependent reduction in the efficacy of neuronal synapses.

Long Term Potentiation: Activity-dependent increase in the efficacy of neuronal synapses.

Oja's learning: Multiplicative normalized Hebbian learning.

Spike Timing Dependent Plasticity: Spike-tailored Hebbian-based learning in which the relative timing of pre- and postsynaptic spikes are used to modulate synaptic connection strength.

7.6 FURTHER READING

- **Section 7.1**
  - A review of neural circuits which were shown to link internal activity states to social behaviour in small animals is given in [17].
  - A review of common neural circuit disruptions in cognitive control across psychiatric disorders is given in [198].
- **Section 7.2**
  - The latest research on large scale simulations of detailed neural networks originated from the Blue Brain Project is published on the project's website: www.epfl.ch.
  - Read more about the TrueNorth project in [72] and its utilization for simulating biological models in [181].

  - Read about the utilization of the SpiNNaker for real-time cortical simulations in [244].
  - Cell atlas for the mouse brain [89] is available at bbp.epfl.ch.

- **Section 7.4**
  - A review of STDP and its variants is available in [62].
  - Read about the connections between STDP and BCM in [143].
  - An online course with a detailed Python implementation of STDP: "Computational modeling of neuronal plasticity," is available at training.incf.org. The course is offered by the online hub TrainingSpace.
  - Read about the implementation of STDP in memristors in [256].

III

The Computer Architect's Perspective

CHAPTER 8

Neuromorphic hardware

Abstract

From the computer architect's perspective, neuromorphic hardware has the potential to shed light on some of the greatest challenges in chip design: thermal management and memory wall. Neuromorphic systems emulate the organization and function of nervous systems. They are comprised of analog and digital electronic circuits, fabricated using the Complementary Metal-Oxide-Semiconductor (CMOS) technology using Very Large-Scale Integration (VLSI). These neuromorphic systems directly embody the physical processes that underlie the computations of neural systems in their CMOS circuits. Neuromorphic hardware offers a method of exploring neural computation in a medium whose physical behavior is analogous to biological nervous systems and operates in real-time, irrespective of size. This chapter will review the most basic building blocks in neuromorphic electronics and the design of Silicon neurons (SiN).

8.1 TRANSISTORS AND MICRO-POWER CIRCUITRY

Pure *silicon* is a semiconductor. It conducts electric currents better than insulators but not as well as metal. Pure silicon has four electrons in its outermost shell, allowing it to create bonds with four other silicon atoms, forming a tetrahedral crystal. A silicon crystal lattice comprises a repeating pattern of eight atoms. While conductive, only a small number of electrons has enough energy to move through the lattice. Introducing or *doping* the lattice with *Boron* atoms which have three valence electrons increases the conductance of the lattice, creating a flow of "holes." Boron-based doping is referred to as *p-type doping*, as the Boron atoms are treated as positive particles. Doping the lattice with *Phosphorous*

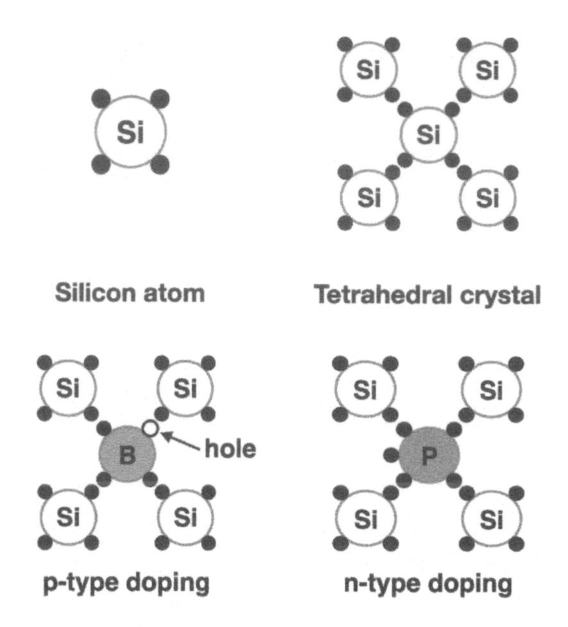

Figure 8.1 Pure silicon has 4 electrons in its outermost shell; it can create bonds with its four nearest neighbors, forming a tetrahedral crystal. Boron (green) doping introduces three valence electrons atoms to the lattice (creating a "hole"; p-type doping) and Phosphorous doping introduces five valence electrons atoms to the the lattice (n-type doping).

atoms, which have five valence electrons, also increases the lattice conductance. Phosphorous-based doping is referred to as *n-type-doping.* Silicon conformations are shown in **Figure 8.1**. When p-type silicon is in contact with n-type silicon (creating a *np junction*), free electrons in the n-side of the junction diffuse into the p-side, creating a *depletion layer*, an insulating region within conductive silicon.

Metal–Oxide–Semiconductor Field-Effect Transistor (MOSFET) is a specialized device in which p-and n-type silicon patterns are fabricated to implement a logical function. A MOSFET has three connections: the *source*, *gate*, and *drain* (A MOSFET's body is usually connected to the source terminal). An *NMOS transistor* is made by using n-type silicon for the source and drain, with p-type silicon placed in-between (in a *PMOS transistor*, it is the opposite). The gate is positioned above the p-type silicon, separated by an insulating layer of silicon dioxide. Normally, there is no current flow between an n-type and a p-type silicon; thus, electrons do not flow between the source and the drain (due to the

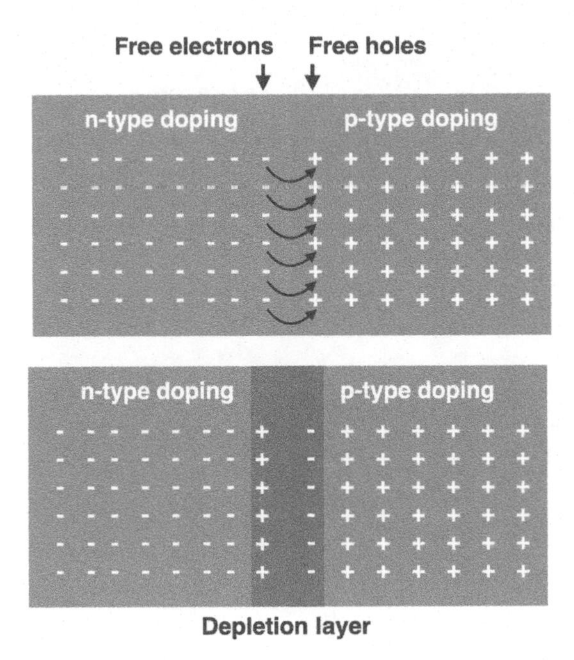

Figure 8.2 In a np junction, electrons diffuse from the n-side to the p-side of the junction, creating an insulating region termed a depletion layer.

formation of a depletion layer). When a positive voltage applied on the gate, the gate electrode creates an electrical field that attracts electrons to the p-type silicon between the source and the drain. That, in turn, changes the area to behave as if it is an n-type silicon (*inversion layer*), creating a path for electrons to flow (inversion charge) and thus, turning the transistor "on" (**Figure 8.3**).

The *strong inversion* MOSFET model assumes that inversion charge goes to zero when the gate voltage drops below a threshold voltage. However, a *weak-inversion* current (Femto to nano amperes) is thermally-driven through the transistor, constituting a *sub-threshold leakage*. Electrons' thermal energy follows the *Fermi–Dirac distribution* of electron energies, allowing some of the more energetic electrons at the drain to enter the channel and pass to the source. Sub-threshold current is exponentially correlated to the gate-source voltage V_{GS}. In micro-power analog circuits, weak inversion is an efficient operating region, and a sub-threshold current is a useful transistor mode around which circuit functions can be designed.

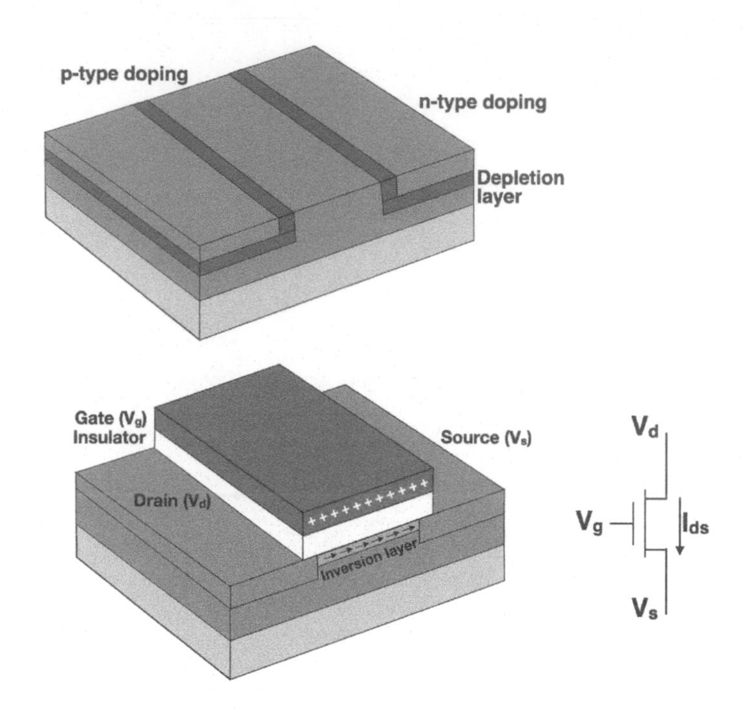

Figure 8.3 The NMOS transistor comprises a source, a gate, and a drain. It has n-type silicon for the source and drain, separated by p-type silicon, above which the gate is positioned. A depletion layer is created at the n-type and p-type interface, blocking the current flow through the transistor. The depletion layer can temporarily be reversed through the induction of an inversion layer, created by applying voltage on the transistor's gate.

NMOS sub-threshold transfer function is:

$$I_{ds} = I_0 e^{k_n \frac{V_g}{U_T}} \left(e^{-\frac{V_s}{U_T}} - e^{-\frac{V_d}{U_T}}\right), \tag{8.1}$$

where I_0 denotes the NMOS current-scaling parameter, k_n is the NMOS sub-threshold slope factor, U_T is the thermal voltage, V_g, V_s, and V_d are the gate, source and drain voltages, respectively. The current is defined to be positive if it flows from the drain to the source.

Eq. 8.1 is equivalent to:

$$\begin{aligned} I_{ds} &= I_0 e^{k_n \frac{V_g}{U_T} - \frac{V_s}{U_T}} - I_0 e^{k_n \frac{V_g}{U_T} - \frac{V_d}{U_T}} \\ I_{ds} &= I_f - I_r. \end{aligned} \tag{8.2}$$

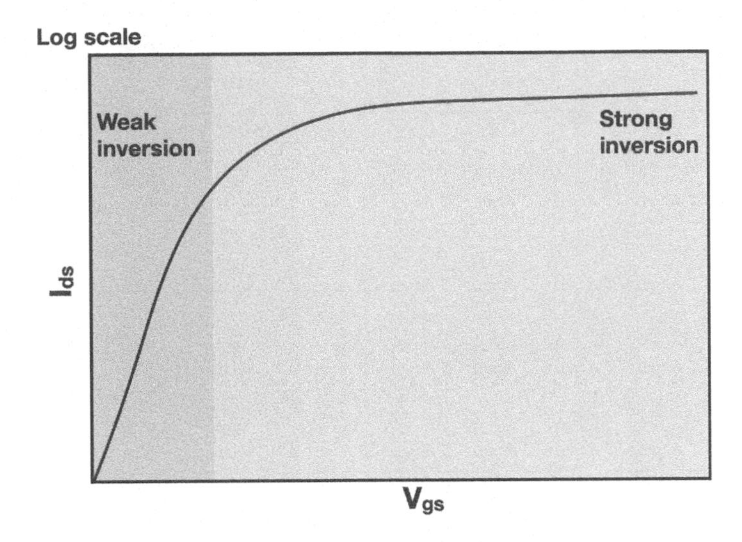

Figure 8.4 In the sub-threshold regime, I_{ds} is exponentially related to V_{gs} (weak inversion regime).

If $V_{ds} > 4U_T$, I_r is negligible and the transistor can be described using:

$$I_{ds} = I_0 e^{k_n \frac{V_g}{U_T} - \frac{V_s}{U_T}} \tag{8.3}$$

In the sub-threshold regime, transistor current (I_{ds}) is exponentially related to V_{gs}. As V_{gs} increases, I_{ds} rises exponentially toward saturation (**Figure 8.4**). When a transistor is operated in the sub-threshold regime, electron transport is governed by diffusion. Therefore, electron transport in this regime follows the same time scales and dynamic as ions do (synaptic time constants of tens to hundreds of milliseconds; ion current increases exponentially with voltage when driven through channels across a cell membrane; see Chapter 5) [201].

VLSI is the process of creating an Integrated Circuit (IC) by combining many thousands of transistors into a single chip. VLSI allows the integration of processing units, memory, communication logic, and other circuitry into one chip. Chip design process is illustrated in **Figure 8.5** and described in length in [173]. VLSI design is a highly specialized and expensive process, comprises computer-aided optimized logical design, components' placing, layout planning, and packaging. The neuromorphic systems, previously described in Section 2.3.4.3 are VLSI circuits.

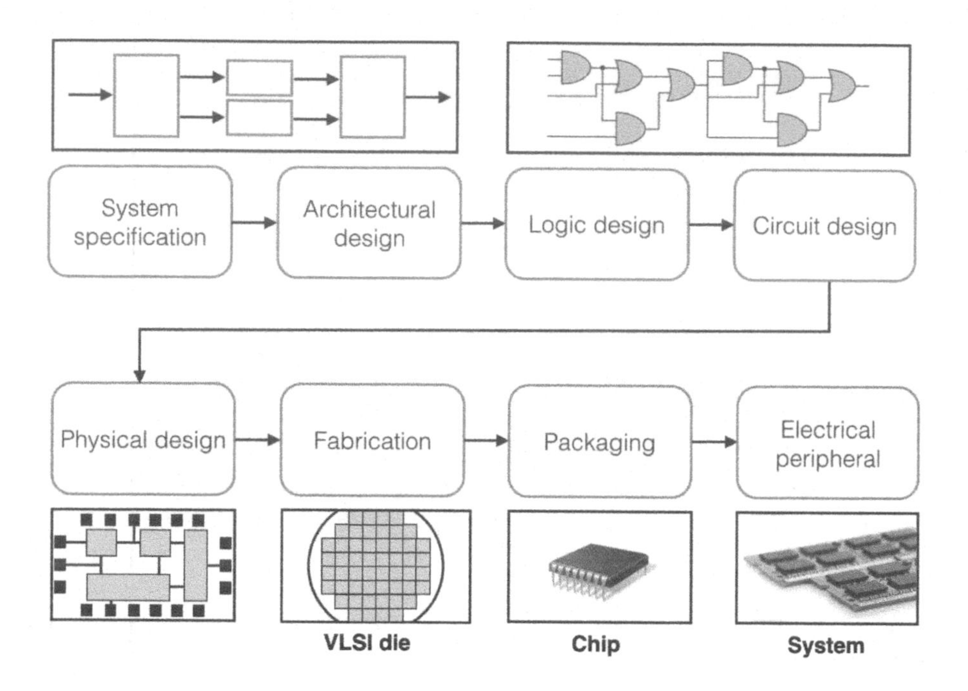

Figure 8.5 The design process of an integrated circuit is usually initiated with system specification within feasibility and functional analysis are performed. Following an initial analysis, system-level architecture is devised, where the main components of the circuit are characterized along with their interfaces, data exchange, and communication protocols. Once subsystems are described, logic and circuit design are initiated and followed by simulation and verification. In the physical design stage, fabrication design takes place - transistors are placed and routed, and timing analysis and layout optimization take place. Finally, the chip is fabricated, packaged, and embedded within a larger board which supplies it with the required peripheral circuits (e.g., voltage stabilization and signal preprocessing).

8.2 THE SILICON NEURON

SiN are circuits that emulate the electrophysiological behavior of biological neurons. They represent one of the main building blocks of neuromorphic electronics. SiNs operate in real-time and communicate, ideally, with scale-independent speed. They offer a medium in which neuronal networks can be emulated directly in hardware rather than simulated on a general-purpose computer, thus providing a qualitative approximation to neurons' actual performance. Many different types of SiNs have been proposed, from complex biophysical-like circuits that emulate an ion channel dynamic and detailed morphologies to basic LIF-based circuits. The basic building blocks of the SiN are:

1. Synapse. A circuit that integrates incoming spikes (changes in voltage) and generates a corresponding current. Usually realized with analog circuitry in which actuated transistors regulate electrical current.

2. Spike generator. A circuit that receives synaptic current and generates a spike of voltage when a certain threshold is reached. Usually realized with a comparator amplifier and some mechanism for voltage reset.

In the following few sections, we will explore a few representative examples of SiN designs. Some of the circuits described below were analyzed by *Chiara Bartolozzi* and *Giacomo Indiveri* in [29]. They are also described in length in [138] and [179]. Here we summarize the essential elements of these circuits. Circuit simulations were performed with *LTSpice* by *Analog Devices* and are available on the book website.

8.2.1 The pulse current-source synapse

The pulse current-source synapse is one of the first synapse circuits, proposed by *Carver Mead* back in 1989 [199]. The circuit schematic is shown in **Figure 8.6**. It is a voltage-controlled current source, activated by a few micro-seconds long active-low input spikes. The resulting current follows the input voltage pattern and dynamic. The current I_{syn} in the pulse current-source synapse is:

$$I_{syn} = I_0 e^{-\frac{k}{U_T}(V_w - V_{dd})} \tag{8.4}$$

where V_{dd} is the power supply voltage, I_0 is the leakage current of a transistor which is activated in the sub-threshold regime, k is the sub-threshold slope factor and U_T is the thermal voltage (at room temperature it is $\approx$ 26mV). This circuit allows the control of the magnitude of the output current (*magnitude control*). When $V_w = V_{dd}$, $Isyn = I_0$, and as we decrease V_w, I_0 is scaled up exponentially, increasing I_{syn} accordingly (**Figure 8.7**). While offering magnitude response control, the pulse current-source synapse does not provide timing nor waveform control.

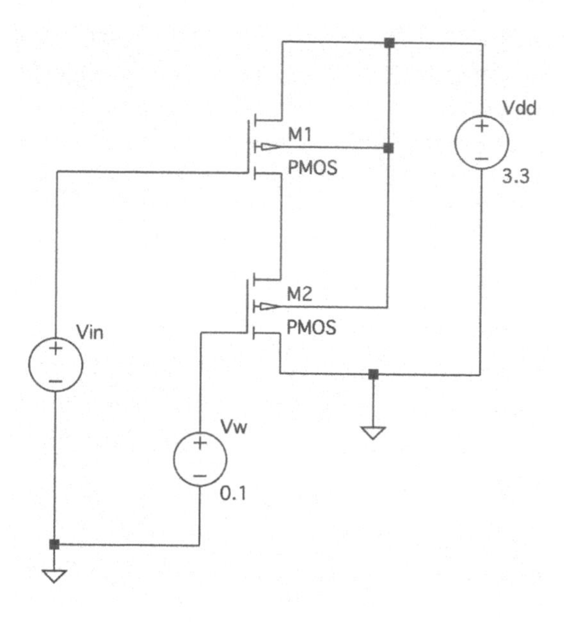

Figure 8.6 The pulse current-source synaptic circuit is comprised of two transistors: $M1$ is working in the saturation regime where $M2$ leakage current is scaled with V_w, offering magnitude control of the output current.

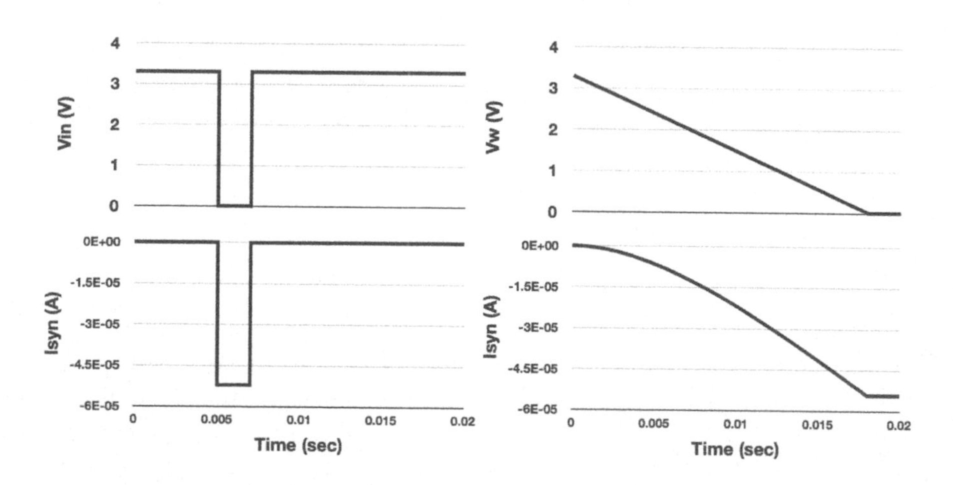

Figure 8.7 When a spike arrives (activated low) as V_{in} to the current-source synaptic circuit, current I_{syn} is generated (left). The circuit offers magnitude control as the level of I_{syn} is proportionally dependent on V_w (right).

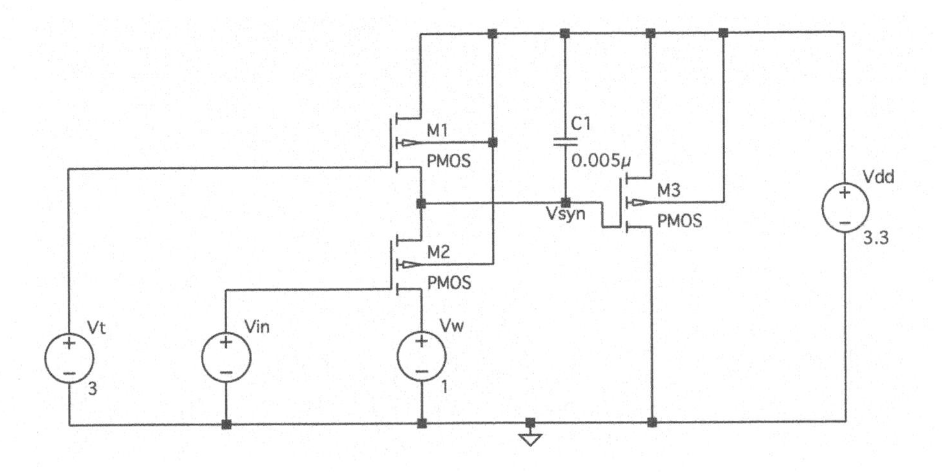

Figure 8.8 The reset-discharge synapse is comprised of three transistors and one capacitor. By controlling the charging rate of the capacitor (through V_t), the circuit provides some temporal control of the generated current's wave form. Similarly to the pulse current-source synapse, V_w provides the current's magnitude control.

8.2.2 The reset and discharge synapse

Described by *John Lazzaro* and *John Wawrzynek* in the 1990s [167], the reset and discharge synapse offer control of I_{syn} signal decay, particularly I_{syn} decay exponentially via the charging and discharging of a capacitor. The circuit schematic is shown in **Figure 8.8**.

Transistor $M2$ is acts as a digital switch, activated by the synapse's incoming spikes, and $M1$ is working in the sub-threshold regime, setting C_1's charging current. $M3$ generates I_{syn}, described with:

$$I_{syn} = I_0 e^{-\frac{k}{U_T}(V_{syn}-V_{dd})} \tag{8.5}$$

For each incoming spike at V_{in}, V_{syn} is driven to the bias voltage V_w. When a spike terminates, M_{pre} is switched off and V_{syn} is linearly driven back to V_{dd}, at a rate set by $\frac{I_\tau}{C_{syn}}$ (**Figure 8.9**), where I_τ is the current driven through transistor $M1$ which is controlled by V_t.

In addition to magnitude control, the reset-discharge synapse provides *temporal control* (V_t regulates the capacitor's charging rate, affecting I_{syn}) (**Figure 8.10**).

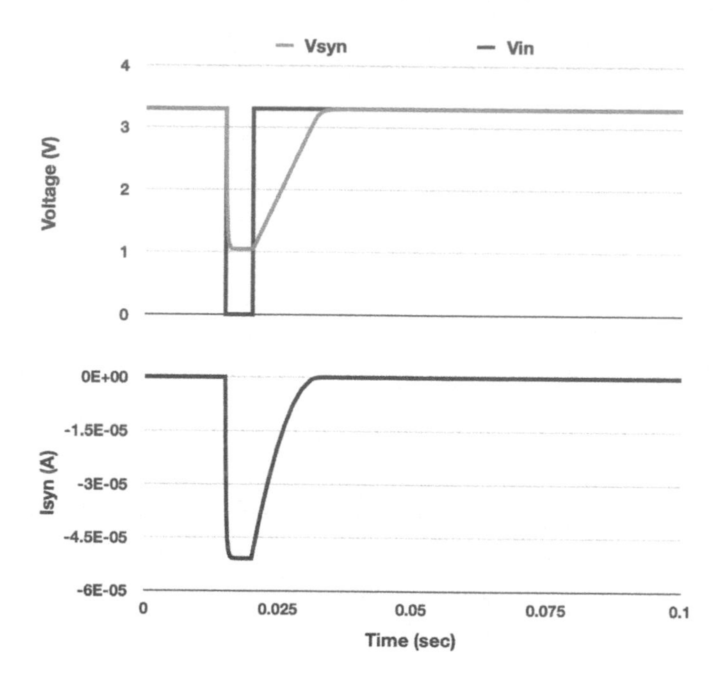

Figure 8.9 Reset-discharge synapse. When a spike arrives (activated low) as V_{in}, current I_{syn} is instantaneously generated. However, in contrast to the pulse current-source synapse, when the spike terminates, I_{syn} slowly falls down at the rate dictated by V_t.

8.2.3 The charge and discharge synapse

The charge and discharge synapse [29] provide temporal control of both charging and discharging of a capacitor, allowing for temporal integration of incoming spikes. The circuit schematic is shown in **Figure 8.11**. In the charge and discharge synapse, the incoming spikes activate the $M4$ transistor which is active high. During a spike, V_{syn} decreases linearly, at a rate set by $I_w - I_t$, where I_w is the current driven though $M3$ (and regulated by V_w) and I_t is the current driven through $M1$ (and regulated by V_τ). This net current is responsible for the capacitor discharge. The resulted linearly decreasing voltage of V_{syn}, causes I_{syn} to increase exponentially (*pulse phase*). In this phase:

$$I_{syn}^{pulse} = I_{syn}^{-} e^{\frac{t-t_i^-}{t_c}} \tag{8.6}$$

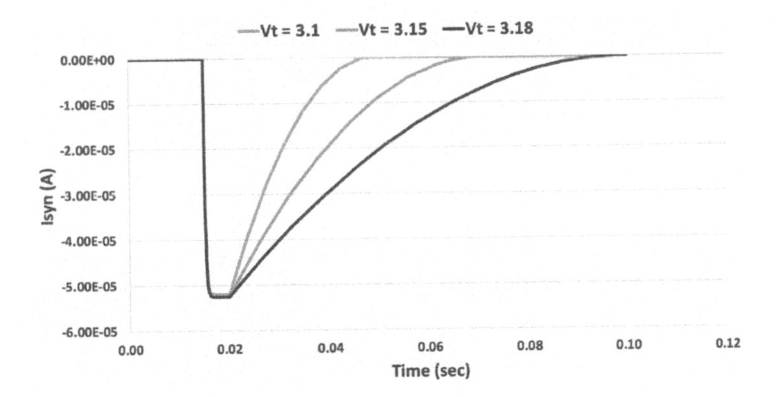

Figure 8.10 Reset-discharge synapse: temporal control. I_{syn} dynamic is dependent on V_t which regulates the capacitor charge cycle. As V_t increases, the charging current I_t decreases and the capacitor charges at a lower rate. As a response, V_{syn} slowly rises, gradually closing the output transistor and I_{syn} gradually decreases.

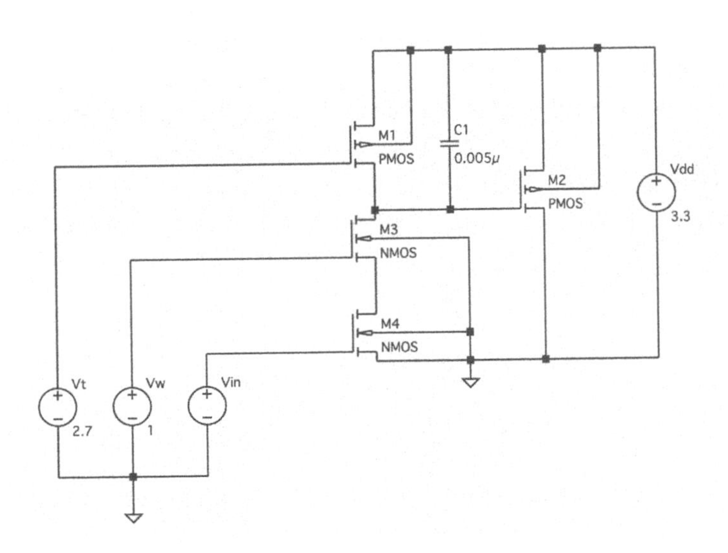

Figure 8.11 The charge-discharge synapse circuit adds another transistor to the reset-discharge synapse, enabling it to control the capacitor discharge (through V_w). This added DOF provides temporal integration of incoming spikes.

where t_i^- is the time the i_{th} spike arrives, I_{syn}^- is I_{syn} at time t_i, and t_c is the time constant for the capacitor charge which equals: $U_T \frac{C}{k(I_w - I_t)}$. In-between spikes, V_{syn} is driven toward V_{dd} at a rate set by I_t. As a result, I_{syn} decreases exponentially (*discharge phase*). In this phase:

$$I_{syn}^{flat} = I_{syn}^+ e^{\frac{t - t_i^+}{t_d}} \tag{8.7}$$

where t_i^+ is the time the i_{th} spike terminates, I_{syn}^+ is I_{syn} at time t_i^+, and t_d is the time constant for the capacitor discharge which equals: $U_T \frac{C}{kI_t}$.

The charge and discharge synapse provide temporal control of both the charging and discharging of a capacitor. Temporal control over the capacitor's charge cycle offers the ability to temporally integrate multiple incoming spikes (**Figure 8.12**). This property is crucial, as synapses can receive spikes from thousands of cells.

The main drawback of the charge-discharge synapse is the fact that it not a linear integrator.

8.2.4 The log-domain integrator synapse

This circuit, proposed by *Paul Merolla* and *Kwabena Boahen* [206], offers linear integration of incoming spikes. The circuit schematic is shown in **Figure 8.6**.

This circuit exploits the logarithmic relationship between the transistor's V_{gs} voltage and its channel currents to overall exhibit linear properties (**Figure 8.14**).

During the charge phase, the circuit governing equation is:

$$I_{syn}^{pulse} = \frac{I_o I_{w0}}{I_t}(1 - e^{-\frac{t - t_i^-}{\tau}}) + I_{syn} e^{-\frac{t - t_i}{\tau}} \tag{8.8}$$

and during the discharge phase is:

$$I_{syn}^{flat} = I_{syn}^+ e^{-\frac{t - t_i^+}{\tau}} \tag{8.9}$$

Detailed analysis of this circuit is out of the scope of this book; however, the interested reader is advised to refer to [29].

The log-domain integrator is a synaptic circuit with linear filtering properties which allows a linear summation of incoming spikes. Other more advanced implementations offer optimized designs with easier fabrication and support of shorter spikes (this circuit design requires long-enough spikes to supply a sufficient charge for the membrane capacitor). However, the same electrical and computational principles apply.

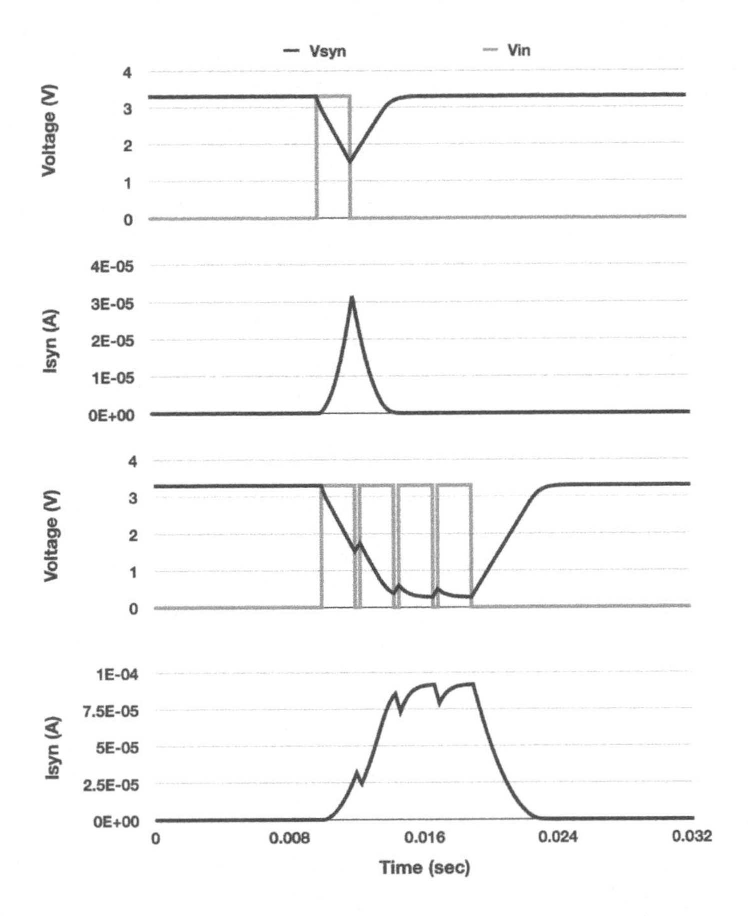

Figure 8.12 The charge-discharge synapse offers control of two temporal properties - the rise and fall times of the generated spike (top). When spikes arriving as V_{in} at the charge-discharge synaptic circuit, they are temporally integrated. Each spike contributes to I_{syn}, prolonging and increasing the generated current, until saturation (bottom).

8.2.5 The axon-hillock neuron

The Axon-Hillock circuit [199] is one of the first circuits proposed for generating LIF-inspired spiking neuron (Section 5.1). It produces a spike when the membrane voltage exceeds a voltage threshold, and it is based on an op-amp comparator. A circuit schematic is shown in **Figure 8.15**.

Input currents ($B1$) are integrated on the membrane's capacitor C_{mem}. As a result, V_{mem} (at the + entrance to the amplifier) linearly increases until it reaches the amplifier switching threshold, defined with

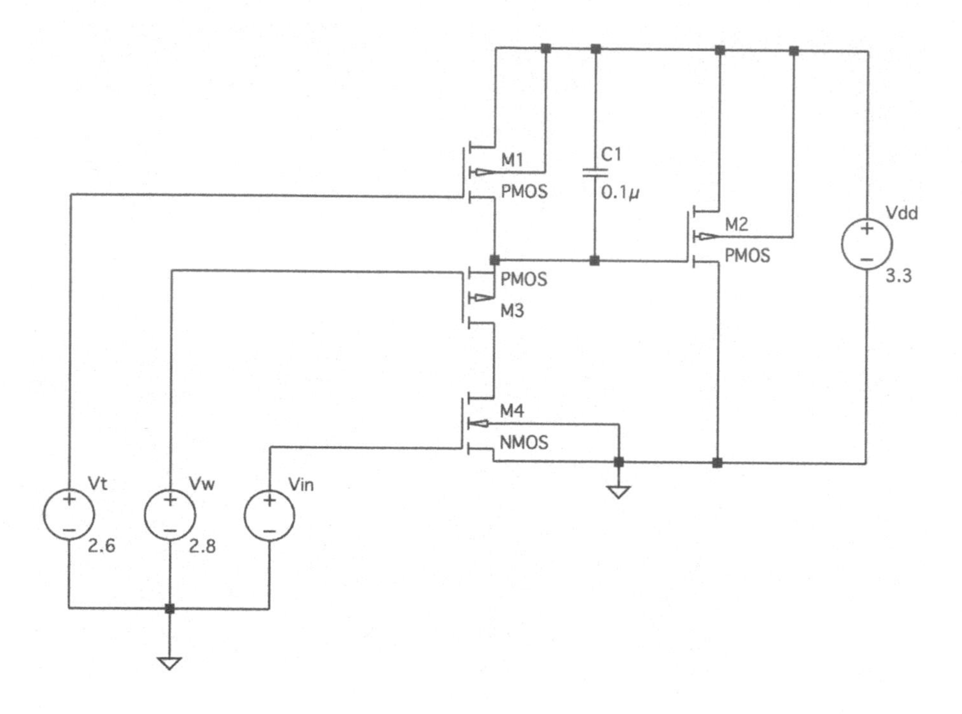

Figure 8.13 The log-domain integrator synapse circuit looks very similar to the charge-discharge circuit, with two differences: the $M3$ transistor is PMOS (instead of NMOS) and its body is connected to V_{syn} (instead of to the ground). These changes allow the circuit to feature linear integration of incoming spikes.

V_{th}. At this voltage crossover, V_{out} (amplifier output) quickly changes from V_s (usually ground) to V_s+ (usually V_{dd}). V_{out} activates the reset transistor ($M7$). A reset current is driven through transistor $M5$ which is regulated by V_{pw}. If the reset current is larger than the input current, the membrane capacitor is discharged until it reaches the amplifier's switching threshold again. At this point, V_{out} swings back to 0V and the cycle repeats. V_{out} is also connected via positive feedback through a capacitor divider C_{fb}.

In this spike generation circuit, the inter-spike interval is inversely proportional to the input current, and spike duration depends on both the input and reset currents. Simulated circuit behaviour is shown in **Figure 8.16**.

The spikes produced are not biologically plausible, as they are not separated by a refractory period and do not exhibit the dynamic and

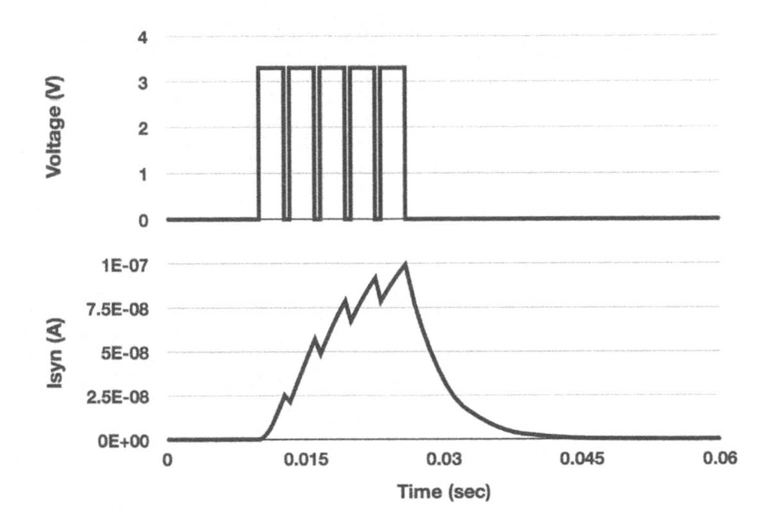

Figure 8.14 The log-domain integrator synapse features a linear integration of incoming spikes where ahead of saturation, each spike equally contributes to I_{syn}.

shape of their biological counterparts. Refractory periods are essential, as they define an upper boundary for the spiking rate.

8.2.6 Voltage-amplifier LIF neuron

Proposed by *Andre Van Schaik* [289], the voltage-amplifier LIF neuron enhances the axon-hillock neuron with more detailed control of the generated spikes, including the spike's rise time, width, fall time, and refractory period. The circuit schematic is shown in **Figure 8.17**.

The capacitor C_{mem} models the neuron membrane and V_{lk} which regulates the conductance of transistor M_{lk} and controls its leakage current I_{lk}. In the absence of input current from incoming spikes (flat phase), I_{lk} drives the membrane voltage V_{mem} (at the + entrance to the amplifier) to 0 V (ground). When input current is apparent, the net incoming current $(I_{in} - I_{lk})$ charges C_{mem}, increasing V_{mem} accordingly. Similar to the axon-hillock neuron, when V_{mem} exceeds V_{th}, an action potential is generated. The action potential is introduced to a voltage inverter, where high (logical) voltage values are transformed to low voltage values and vice-versa. The inverted low voltage activates transistor M_{na}, through which I_{Na} current is driven, charging C_{mem} and creating a sustained high output voltage (constituting the spike). This process directly

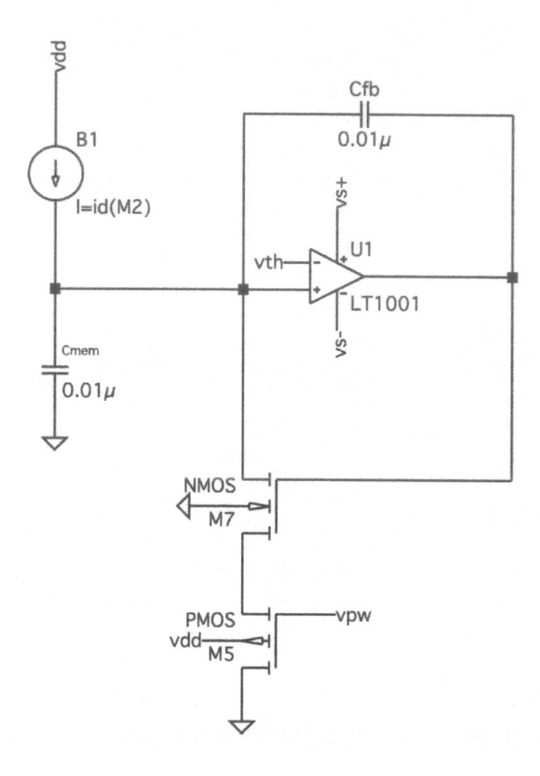

Figure 8.15 The axon hillock circuit is comprised of a capacitor C_{mem} which integrates the incoming current, a comperator and two feedback loops. One feedback loop, through C_{fb}, creates positive feedback for spike generation and the second, activates a discharging reset current. V_{syn} decreases as long as the incoming current is smaller than the reset current.

embodies biological behavior in which an influx of sodium ions (N_a^+) depolarizes the cell membrane during the initiation of an action potential (Section 4.1). In biology, a delayed outflux of potassium ions (K^+) repolarizes the cell. Here, a second voltage inverter drives I_K currents with a similar aim. Through the second voltage inverter, current I_{Kup} charges capacitor C_k. As the voltage on C_k increases, it activates transistor M_K through which current I_{kdn} discharges C_{mem}. Similar to the axon-hillock neuron, when V_{mem} drops below V_{th}, the amplifier output drops as well. As a response, the first voltage inverter takes the voltage up, deactivating M_{na} and terminating I_{NA}. The second voltage inverter takes the voltage down, terminating I_{Kup} and allowing C_k to be discharged by

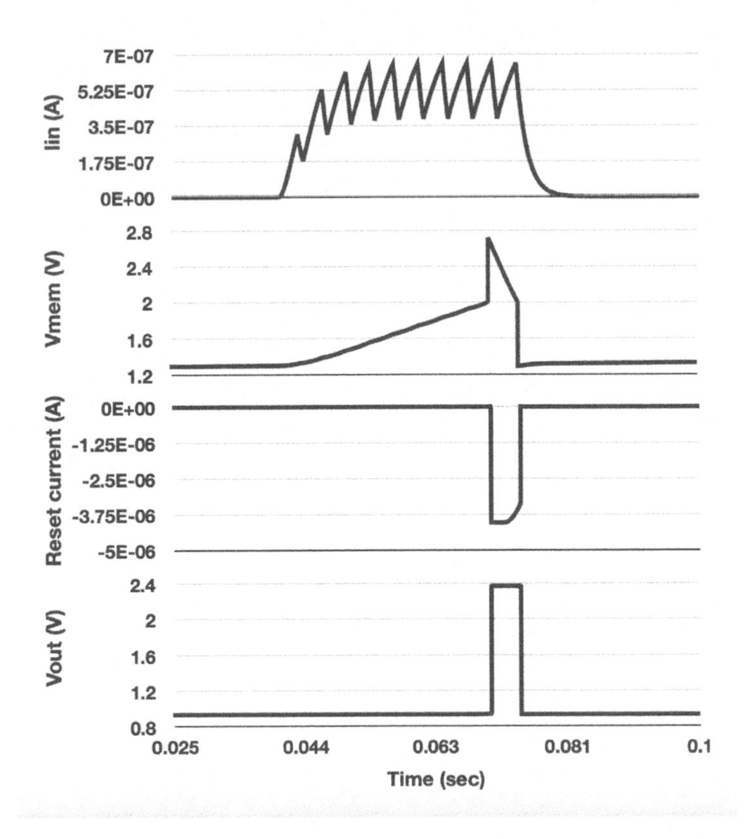

Figure 8.16 Axon hillock: wave traces. The neuron is driven by I_{in}, generated by the charge-discharge synapse (as was described in Section 8.2.3). I_{in} linearly increases V_{mem} (through the charging of C_{mem}), until the comperator threshold is reached and a spike is initiated. The initiated spike induces a reset current which decreases V_{mem} and terminates the spike.

current I_{ref}. As long as I_{ref} has not sufficiently discharged C_k, the circuit cannot be further stimulated by an incoming current (constituting a refractory period). This process is shown in Figure 8.18.

8.3 CASE STUDY: HARDWARE AND SOFTWARE CO-SYNTHESIS

This case study is based on our neuromorphic design, named OZ, of a spiking neuron which can be modulated in accordance with NEF [112]. NEF was mentioned in Section 1.2.2 and will be described in detail in Chapter 12. The OZ neuron has the advantage of being adjustable in

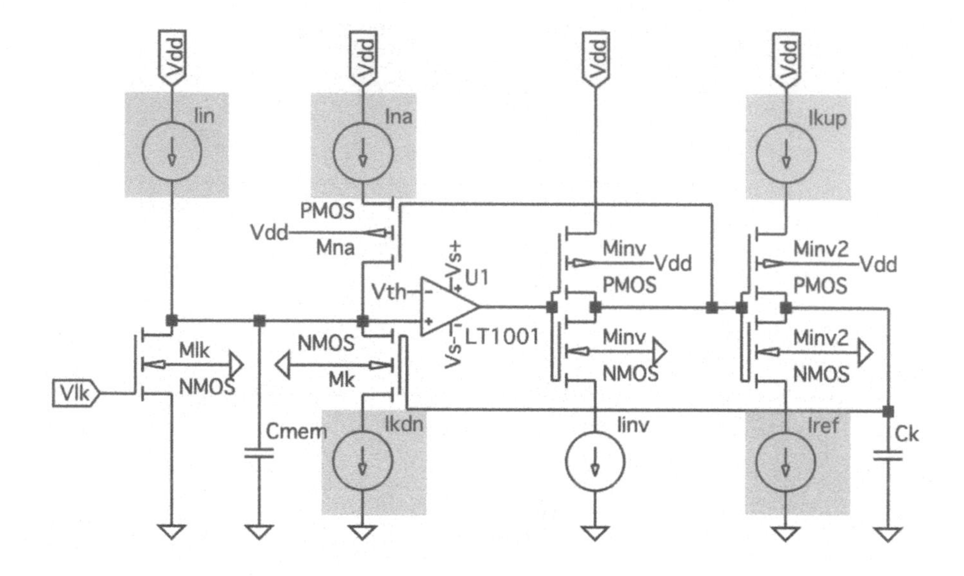

Figure 8.17 The voltage amplifier LIF neuron adds two voltage inverters to the axon-hillock circuit. The first inverter activates I_{NA} through M_{na} and the second activates the I_k currents. Together they simulate the behavior of a biological neuron, providing control of the wave form of the generated spike.

terms of its intercept, the value for which the neuron starts to produce spikes at a high rate, and its maximal firing rate. NEF's neurons are also characterized by encoders, defining the preferred stimulus for that neuron. Examples of six OZ neurons, each configured to feature a different intercept and a maximal firing rate, three negatively encoded and three positively encoded, are demonstrated in **Figure 8.19**. Each neuron was stimulated with the same input voltage which linearly increased from −1 to 1 V over one second.

8.3.1 Circuit design

In the OZ circuit design, stimulus x is introduced through preprocessing modules to two branches, one connected to positively encoded OZ neurons and the other to negatively encoded OZ neurons. These preprocessing modules accept an input voltage ranging from −1 to 1 V (corresponding to the default input normalization scheme taken by NEF) and produce an output voltage ranging from 0 to 3.3 V. Each OZ neuron comprises two consecutive modules:

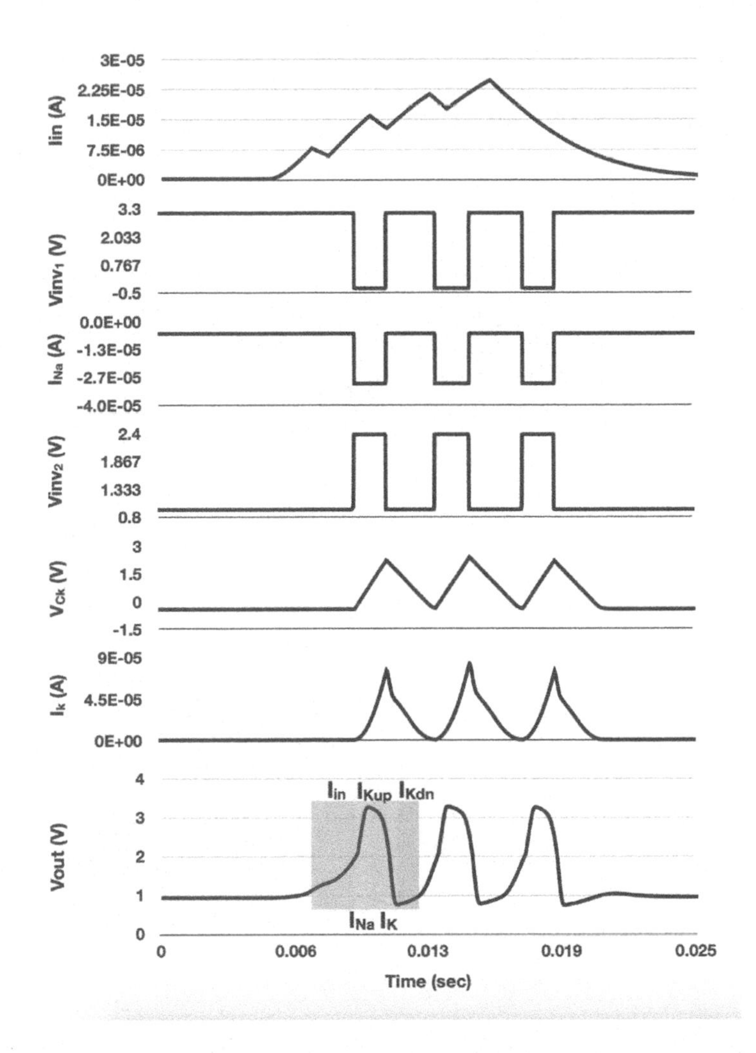

Figure 8.18 Voltage amplifier LIF neuron: wave traces. The neuron is driven by I_{in}, generated by the log-domain integrator synapse (as was described in Section 8.2.4). The current is integrated until the comperator threshold is reached and the two inverters V_{inv1} and V_{inv2} generate I_{NA} and I_k respectively. I_k is activated by C_k. The interplay of the involved current creates biological plausible spikes, including the support of a refractory period.

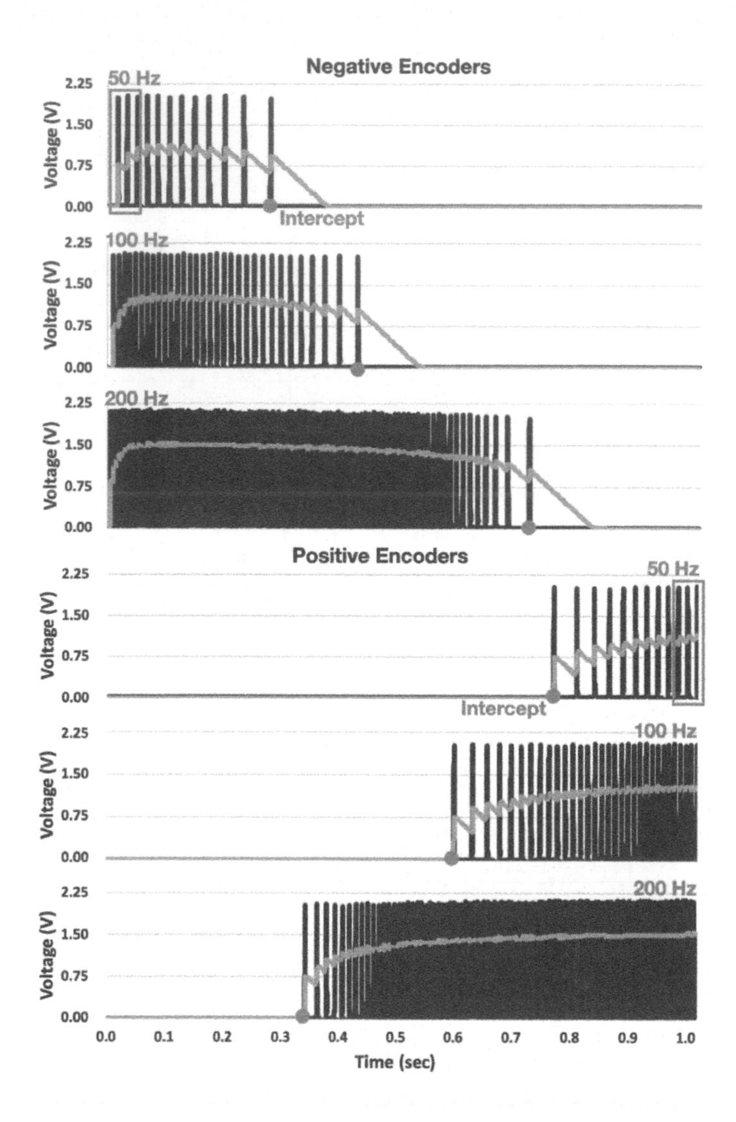

Figure 8.19 Six OZ neurons, three negatively encoded and three positively encoded, each with a different intercept and a maximal firing rate.

1. Spike generator. Characterized by a tuning curve, modulated using control signals, thus realizing NEF-based encoding (to be discussed in Chapter 12).

2. Temporal integrator. The generated spike train is introduced to a temporal integrator, integrating the incoming spikes, producing a continuous time-varying signal.

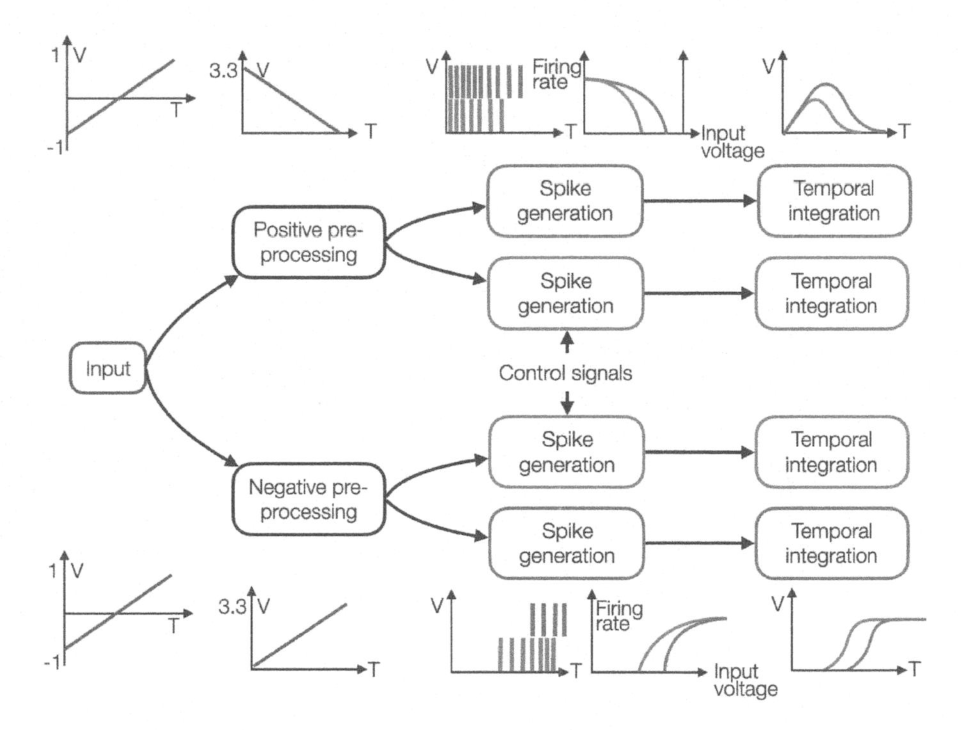

Figure 8.20 Schematic of four OZ neurons grouped into two branches, one for positive and the other for negative encoders. Each branch initiates with an input preprocessing module and each neuron comprises a spike generation and temporal integration circuits. Each neuron has its own response dynamic, and it therefore produces spikes at different times and rates.

The circuit schematic for two negatively encoded and two positively encoded neurons is shown in **Figure 8.20**.

The negative preprocessing circuit comprises two consecutive modules: the first one inverses the voltage and aligns it to initiates at 0 V, and the second reinverts and scales it so that it will terminate at 3.3 V (circuit's V_{dd}). The first module uses an op-amp based adder to add 1 V to the input signal (aligning it to 0 V) and inverts it according to V_o=-(V^++V^-), where V^+ and V^- are the op-amp's input terminals. The resulted voltage ranges from 0 to 2V. The second module uses an inverter amplifier which scales its input voltage according to $-R_{fb}/R_{in}$, where R_{fb} is the feedback resistor and R_{in} is the amplifier's input terminal resistor. Here, the amplifier is tuned to achieve a scaling factor

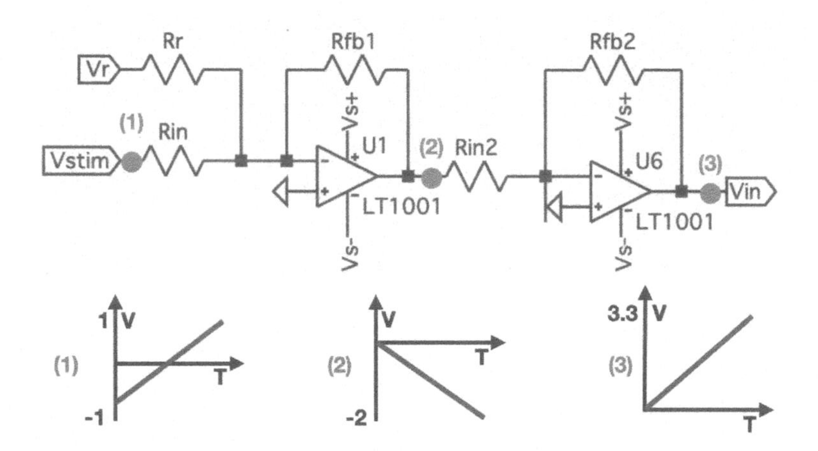

Figure 8.21 Pre-processing module for negatively encoded OZ neurons. The module inverts input voltage, aligns it to start at 0 V, and then re-inverts and scales it so it will start with 0 V and terminate with 3.3 V.

of -1.65 which transforms 2 to 3.3 V. The positive preprocessing module resembles the negative preprocessing module, with the addition of another voltage inverter which produces a similar waveform, initiating at 3.3 V and terminating at 0V. This preprocessing module is shown in **Figure 8.21**.

The OZ neuron's spike generation circuit is shown in **Figure 8.22**. The circuit is based on modified versions of the pulse current source synaptic circuit (for weighted input) (described in Section 8.2.1), the voltage-amplifier LIF neuron (for spike generation) (described in Section 8.2.6), and the sub-threshold first-order low pass filter circuit (for temporal integration) (described in Section 8.2.4). The pulse-current source synapse is used to convert an input voltage to a proportional current, introduced into the spike generation circuit and defined by V_w. The voltage-amplifier LIF neuron's response dynamic is predominantly determined by the values of I_{kup}, I_{kdn}, I_{Na}, I_{ref} and the leakage current I_{lk} (driven through transistor M_{lk}) which are regulated, respectively, by V_{kup}, V_{kdn}, V_{Na}, V_{ref} and V_{lk} via dedicated transistors. Therefore, this neuron has five DOF: V_{lk} controls the discharge rate of C_{mem}, V_{ref} controls the spikes' refractory period, V_{kup} and V_{kdn} control the fall time of the generated spikes and V_{Na} controls the spikes' rise time.

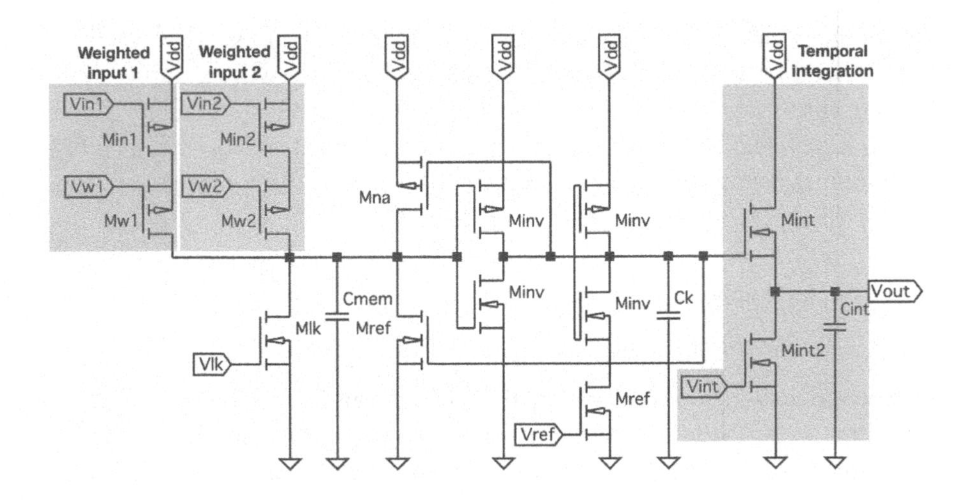

Figure 8.22 OZ's spike generator circuit. Input voltage is transformed to a proportional current which is injected into the voltage-amplifier LIF neuron. The neuron produces a spike train according to its response dynamic. Finally, the spike train is introduced into an integrator circuit to provide temporal integration.

Furthermore, its spiking dynamic relies on an op-amp which is comprised of multiple transistors and resistors. OZ's spike generator is a NEF-optimized circuit design, where V_{up}, V_{Na} and V_{kdn} are redundant. Furthermore, it does not rely on op-amp for spike generation, as the amplifier has no significant functional effect in terms of the neuron's firing rate and intercept (see below). A NEF-tailored design should also enable high dimensional input representation which can be achieved by concatenating input modules (highlighted in **Figure 8.22** as weighted inputs). Further details are given in the circuit analysis section below. Finally, temporal integration can be achieved via a simplified low pass filter. In OZ, capacitor C_{int} is charged by current I_{int} which is activated by the generated spike train and driven through transistor M_{int}. C_{int} discharges at a constant rate through a leakage current which is driven through transistor M_{int2} and regulated by a continuously held voltage V_{int}. The voltage on C_{int} constitutes the OZ neuron's output.

A useful way of representing a neuron's response to varying inputs is by using a response or a tuning curve which is one of the most fundamental concepts of NEF. In NEF, a tuning curve is defined using an intercept, the value for which the neuron starts to produce spikes at a high rate and at its maximal firing rate. OZ's tuning curve can be programmed to control both. For circuit analysis, we will examine eight OZ neurons, four with positive and four with negative encoders. Each neuron has $d+2$ DOF, where d is the dimensionality of the input, corresponding to d values of V_w which regulate each input dimension and V_{lk} and V_{ref} correspond to the two other DOF.

8.3.2 Circuit analysis

8.3.2.1 Architectural design

For the sake of discussion, we will consider one-dimensional OZ neurons. First, we shall consider the classic voltage-amplifier LIF neuron, described in Section 8.2.6. This design relies on an op-amp for spike generation. From a functional perspective, the op-amp provides the neuron with a digital attribute, splitting the neuron into an analog pre-op-amp circuit and a digital post-op-amp circuit. When an incoming current induces V_{mem} to exceed a predefined threshold voltage, the op-amp yields a square signal which generates a sharp I_{Na} response. This fast response induces sharp swing-ups in V_{mem} and V_{out}. Without the op-amp, this transition between states is gradual. While both designs permit spike generation, the op-amp-based design can generate spikes at a higher frequency and amplitude. To compensate for that, we can discard both I_{Na} and I_{kup} controls by removing their regulating transistors. Removing these resistance-inducing transistors maximizes I_{Na} and I_{kup}, thus achieving op-amp-like frequency and amplitude. Moreover, without the op-amp, there is no need to explicitly define a threshold, providing a more straightforward and biologically plausible design.

8.3.2.2 Neuron control

Did we lose control over our neuron's maximal firing rate by eliminating the regulation of I_{kup}? Fortunately, both I_{kup} and I_{ref} impact neuron's firing rate. While I_{kup} limits neuron's firing rate by governing the rise time of the generated spikes, I_{ref} does that by setting the refractory period between spikes. Controlling both currents is redundant as both imply similar constraints, as shown in **Figure 8.23, A** (V_{lk} and V_w were held constant).

V_{lk} controls the discharge rate of C_{mem} by regulating a leakage current through transistor M_{lk}. As long as this leakage current is lower than the input current (driven through the weighted input module), V_{mem} will

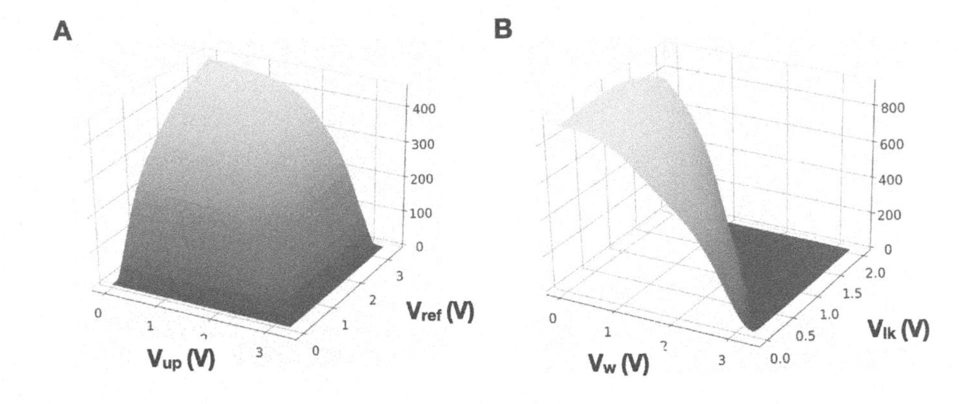

Figure 8.23 A. As both V_{lk} and V_w determine the charging rate of the membrane capacitance, they constrain each other in regard to the neuron's firing rate; B. Neuron's firing rate as a function of V_{up} and V_{ref}.

rise towards saturation. As we decrease V_{lk}, leakage current drops and C_{mem} charges faster. As a result, the neuron's intercept increases for positively encoded neurons and decreases for negatively encoded neurons. Neurons exhibit a maximal firing rate when their input voltage is either −1 V or 1 V, depending on the neuron's encodings. The maximal firing rate is proportionally dependent on the charging status of membrane capacitance C_{mem}. The faster C_{mem} charges, the more frequently the neuron emit spikes. However, a neuron's maximal firing rate is not entirely decoupled from its intercept. While a neuron's intercept is controlled by V_{lk}, it can also be modulated by V_w which provides a magnitude control for the input current. Therefore, while V_{lk} can be used to define the neuron's intercept, V_w can impose on it a firing rate constraint. For example, the neuron's spiking rate will not exceed 400 Hz when V_w is set to 2.2 V. V_{lk} and V_w imposed constraint on a neuron's spiking rate is demonstrated in **Figure 8.23, B** (V_{ref} and V_{up} were held constant).

Figure 8.24 summarizes the control of the neuron's intercept and spiking rate using V_{ref} and V_{lk} (while V_w were held constant).

8.3.3 NEF-inspired design

In **Figure 8.25**, the tuning curves of our eight OZ neurons, along with the tuning curves of eight simulated neurons which were computed directly with NEF are demonstrated. The tuning curves indicate varying

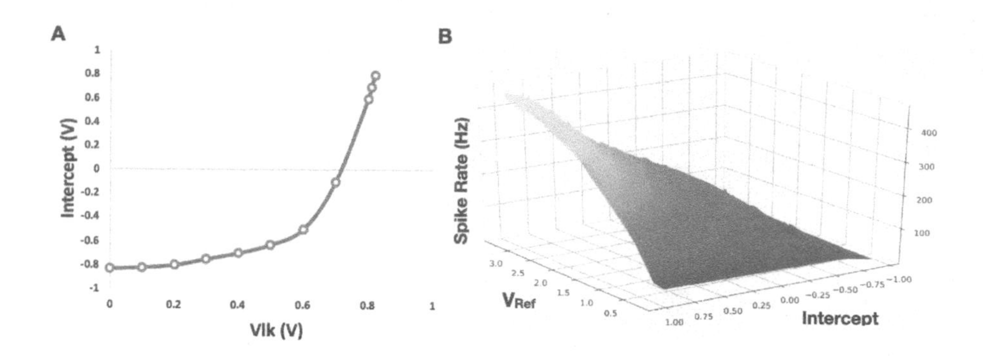

Figure 8.24 A. Neuron's intercept as a function of V_{lk}; B. For a given intercept (determined by V_{lk}), a neuron's spiking rate can be determined by V_{ref}. V_w was held constant in both panels.

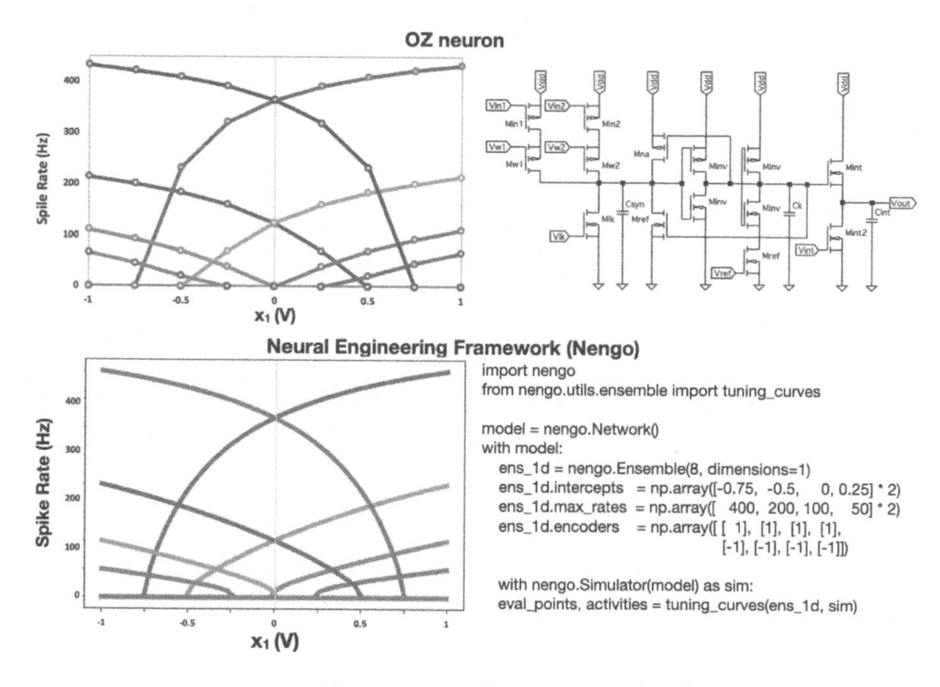

Figure 8.25 Tuning curves of eight 1D hardware-based OZ neurons and NEF-based simulated neurons, demonstrating full correspondence.

intercepts and spiking rates, showcasing the produced spike trains' high predictability and the full correspondence between OZ and NEF.

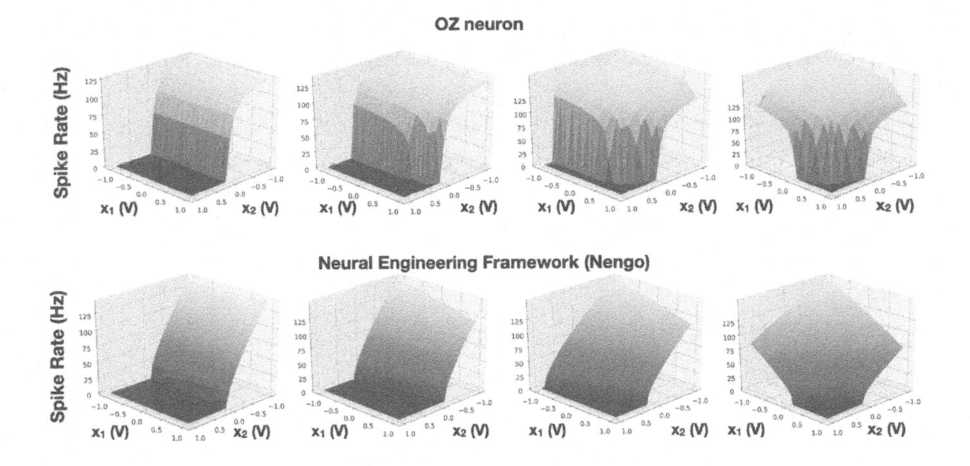

Figure 8.26 The tuning curves of four 2D hardware-based OZ neurons and 2D NEF-based simulated neurons, demonstrating full correspondence in high dimension.

In **Figure 8.26**, 2D tuning curves are demonstrated. 2D tuning is achieved with OZ by concatenating two weighted inputs: x_1 and x_2, regulated by V_{w1} and V_{w2}, respectively. Results show the high predictability of the neuron in response to a high dimensional stimulus.

8.4 GLOSSARY

Integrated circuit: An assembly of electronic components, integrated and interconnected onto a chip.

Silicon neuron: An electrical design which follows the electrophysiological behavior of biological neurons.

The axon-hillock neuron: A primary spike generation circuit which offers firing rate tuning.

The charge and discharge synapse: A reset and discharge synapse with an added capacity for signal integration of incoming spikes.

The log-domain integrator synapse: A charge and discharge synapse in which signal integration is held linearly.

The pulse current-source synapse: An electrical micro-power circuit offering magnitude control over spike-induced current.

The reset and discharge synapse: A pulse current-source synapse with an added temporal control over the spike-induced current.

Very large-scale integration: The process of creating an integrated circuit comprises numerous nano-scale transistors onto a chip.

Voltage-amplifier LIF neuron: The axon-hillock neuron with an added capacity for spike's form modulation.

8.5 FURTHER READING

- **Section 8.1**
  - A nice introduction to transistors is available in the classic Scientific American article "The future of the transistor" [155].
  - A formal introduction to transistors and VLSI circuits is the classic introductory book to VLSI, written by *Carver Mead* and *Lynn Conway* in [202].
- **Section 8.2**
  - The classic book discussing neuromorphic VLSI circuits, edited by *Carver Mead* and *Mohammed Ismail* is given in [203].
  - A prominent review of the discussed synaptic circuits is given in [138]. Some of the concepts are also given in the classic paper by *Rodney Douglas*, *Misha Mahowald*, and *Carver Mead*, given in [79].

CHAPTER 9

Communication and hybrid circuit design

Abstract

Although some of the greatest successes in neuromorphic designs are based on analog circuitry, they are mostly limited in scalability and programmability. Hybrid analog/digital architectures aim at solving the scalability bottleneck by handling spike communication digitally. Pure digital systems aim for high-level programmability and on-chip learning. In this chapter, we move forward from refined analog designs to hybrid and digital neuromorphic circuits. In the first part of this chapter, we will cover the essentials of Address Event Representation (AER). In the second part, we will shift gears toward digital neuromorphic designs, focusing on IBM's TrueNorth architecture as a case study.

9.1 COMMUNICATING SILICON NEURONS

Neurons communicate with spikes. Three classes of communication architectures for neuromorphic systems were identified by *Kwabena Boahen* in [42]: fully dedicated, shared axon, and shared synapse. In a fully-dedicated communication architecture, a dedicated physical wire is allocated to each neuron-synapse connection. Therefore, to support a fully connected connectivity scheme in which any neuron can communicate with any other neuron, N neurons require N axons and N^2 synapses (**Figure 9.1**). However, physically connecting SiNs with N^2 synapses, considering the biological scale by which neurons can be connected to thousands of neurons, is not efficient as ICs are fabricated with a

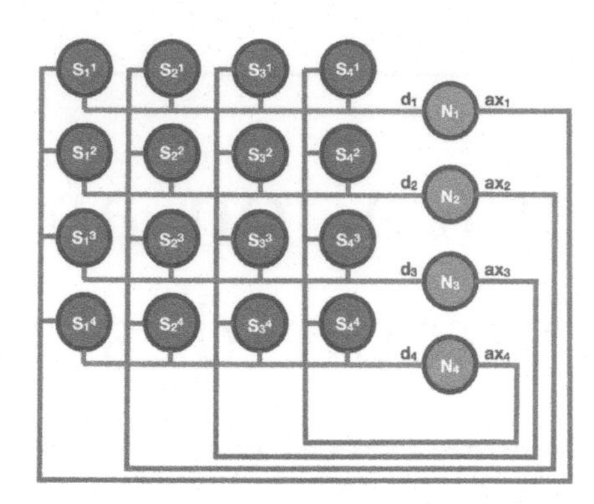

Figure 9.1 In a fully dedicated neuron connectivity scheme, a system with N neurons has N axons and N^2 synapses. Neuron i is indicated with N_i, axons coming out of neuron i is indicated with ax_i and a synapse connecting neuron i with neuron j is indicated with S_{ij}.

limited number of layers and, in each, only planar wiring is possible, constraining design routing.

In the early 1990s, to circumvent these communication limits, a new communication scheme was proposed by *Carver Mead* and colleagues. Instead of allocating a dedicated physical wire to each SiN axon, a wire can be shared among populations of SiNs. This is possible since digital communication allows signals to travel faster than in biological systems. This communication scheme was termed AER. AER communicate spikes as packets of information which hold the necessary data required for spikes routing to target. AER uses a *packet switching* strategy in which communicated packets (or spikes) share physical resources. Each packet can hold a varying degree of information ranging from the spike's time and source to a detailed description of its dynamic. In the basic version of AER, spikes are encoded with the address of the neuron which generated them (**Figure 9.2**). AER encoding is commonly implemented by having neurons arranged in a 2D grid where one encoder encodes the x index and another the y index of the spiking neuron.

The canonical way to implement address encoding is by using a logarithmic encoder in which each address line is connected to log_2N wires to constitute the neuron address. Therefore, by utilizing AER, a shared axon architecture would require log_2N axons to support fully connected N neurons (**Figure 9.3**). Other encoding mechanisms include the hierarchical [98] and counter-based encodings, both summarized in [179].

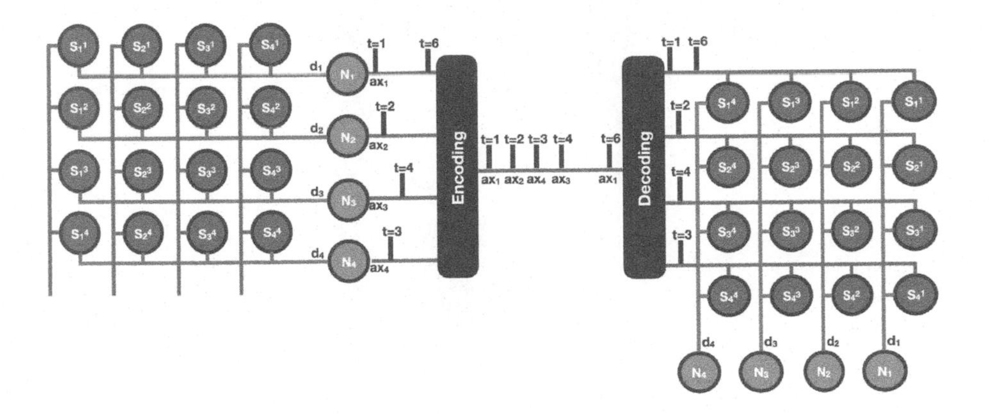

Figure 9.2 With address event representation, spikes are encoded with a neuron address and multiplexed on a communication line by an encoding circuit. Spikes are distributed to their targets by an encoding circuit.

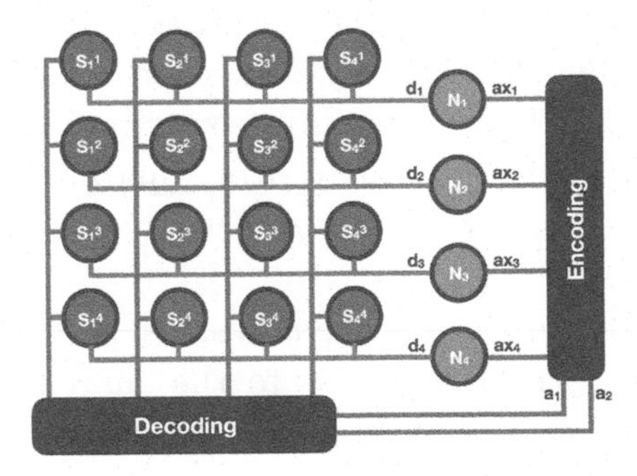

Figure 9.3 In a shared axon neuron connectivity scheme, neuron connectivity is made via an encoder and decoder circuits which implement AER. This architecture features N neurons, log_2N axons and N^2 synapses.

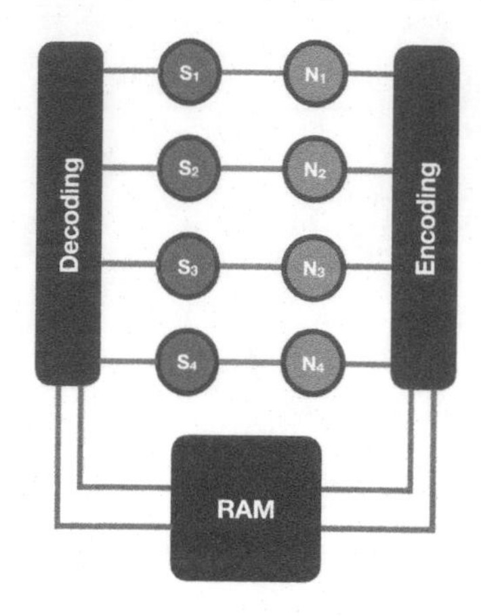

Figure 9.4 In a shared synapse neuron connectivity scheme, neuron connectivity is made via a memory chip which holds a programmable address and weight mapping. This architecture features N neurons, $log_2 N$ axons and N synapses.

The shared axon architecture still has to support N^2 synapses. To alleviate this constraint, the shared synapse architecture uses memory to realize axonal branching. Each shared synapse circuit retrieve the connectivity scheme from memory, using the communication line's address. The shared synapse architecture uses $log_2 N$ axons and N synapses to support fully connected N neurons (**Figure 9.4**). This architecture is particularly appealing for sparsely connected neurons. While a shared axon design features direct mapping (sometimes hard-coded) between neurons, a shared synapse design stores mapping in memory. With memory-based mapping, various connectivity schemes can be defined with address tables.

As with all asynchronous system designs, some arbitration mechanism is utilized to prioritize access to the shared resource. Many arbitration strategies have been suggested including: bus sensing [3], tree arbitration [41], ring arbitration [136], multidimensional arbitration [38], and eliminating arbitration [215]. These approaches are summarized in [179]. Here we will discuss the most widely utilized arbitration method: the tree arbitration. A naive arbitration approach would demand each neuron to be prioritized against all other neurons, constituting an overhead of $O(N)$ operations for each spike. In a tree arbitration scheme, arbitration takes place between neurons, hierarchically connected in a

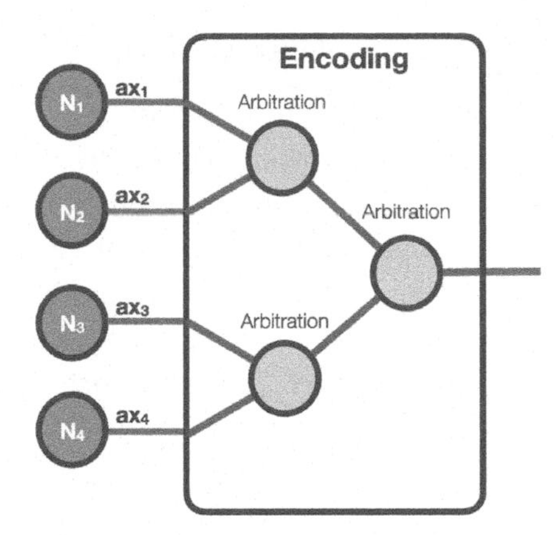

Figure 9.5 In a tree arbitration scheme, spiking neurons compete for a shared communication line via hierarchical binary arbitration units. Arbitration takes an overhead of $O(log_2N)$ operations.

binary tree. The spiking neurons' addresses are held in a queue and communicated forward when the communication line is free. As a result, while spikes are not lost, their timing data might be compromised, as some spikes have to wait for the communication line availability. This has a dramatic effect on the throughput of spike communication with a $O(log_2N)$ overhead of arbitration.

Spike communication imposes an interesting trade-off between the spike's time and data. We can preserve spike data by using arbitration or preserve timing by doing randomized line assignments to deal with collisions naively. Another option is to encode a spike's time in the data packet. This however, requires a more complicated circuitry for packet processing. As with other complicated systems, neuromorphic designs balance requirements, technical limitations, and optimization algorithms (design trade-offs).

Employing analog circuitry for computation and digital circuitry for communication created a hybrid approach toward neuromorphic hardware. It elegantly combines the analog and the digital realms to get the best of both: the analog circuitry's computing efficiency and the digital communication's performance and reliability. Hybrid neuromorphic designs have, therefore, paved the way for widely utilized neuromorphic hardware.

9.2 FROM HYBRID TO DIGITAL CIRCUITRY

Analog and hybrid neuromorphic designs have few main drawbacks: limited correspondence between software (the neural algorithm) and hardware (the analog implementation); limited support of functional neural circuits with biological constraints (e.g., time constants); and scaling constraints due to lack of high-density capacitors in modern CMOS processes. Due to these challenges and others, companies like IBM and Intel have taken a pure digital approach toward neuromorphic hardware. To demonstrate how digital architecture can be designed, IBM's TrueNorth digital architecture will be described next.

A neuromorphic digital architecture is often a semi-synchronous system: while neurons' states are synchronously updated, communication is based on spiking events. In the TrueNorth chip, a discrete neuronal model is described using a synchronous and (not biologically plausible) update rule, where the voltage of neuron i is updated in each time step [207], using:

$$V_i(t+1) = V_i(t) + L_i + \sum_{j=1}^{k} A_j(t) W_{ji} S_{ij} \tag{9.1}$$

where W_{ji} is a binary synaptic variable, indicating if axon j is connected to neuron i, S_{ij} is the synapse weight between axon j and neuron i (ranging between -256 to 256) and L_i is the neruon's leak current (ranging between -256 to 256). When V_i in time t exceeds its threshold value Θ_i, it generates a spike event, and its V_i is reset back to 0 in $t+1$. The spike event is encoded as an address and sent off the core. Negative values of V_i are set to 0 at the end of each time step (implementing a Rectified Linear Unit (ReLU)-like behavior). TrueNorth therefore implements LIF-like neurons (Section 5.1). A basic TrueNorth core has 256 axons, a 256x256 synapse crossbar, 256 neurons, and an update rate of 1 kHz (every 10^{-3} seconds). The synapse crossbar is implemented using a Static Random-Access Memory (SRAM). The crossbar allows for spike communication without sending data off-chip, relaxing the CPU-Memory bottleneck, and appears in von-Neumann computing architecture (Section 2.3). Spikes are communicated from axons (rows in the SRAM memory) to neurons (columns in the SRAM memory) via binary synapses. The activated axon simultaneously communicates with each of the neurons connected to it, thus achieving a highly paralleled communication scheme (**Figure 9.6**). The simple mathematical formulation of Eq. 9.1 and the communication scheme which involve simple

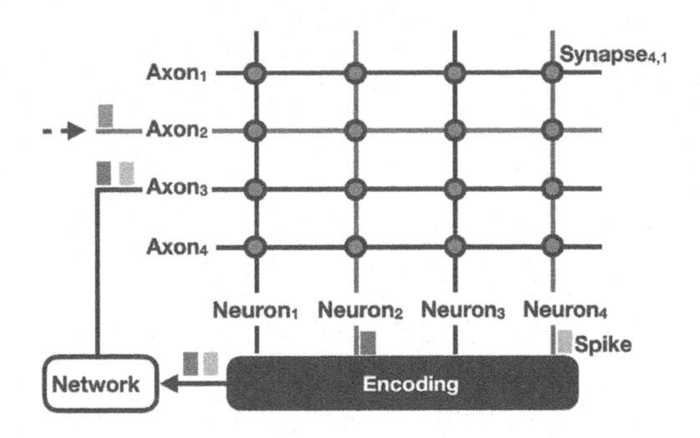

Figure 9.6 A TrueNorth core is comprised of N neurons, connected with N axons through binary synapses. Spikes are encoded with packets and introduced to a router which can route spike events to the same or a foreign core's axons. Here, a spike from a foreign core activates axon 2 which is connected to neurons 2 and 4. As a response, each neuron emits a spike (assuming that the neurons' initial states permit it). In this schematic, both spikes are routed back to axon 3.

addition and multiplication allows for a relatively small circuit design and a limited number of logical gates (avoiding floating-point arithmetic circuitry). A pseudo-random number generator is implemented within a core, providing a source for noise to achieve stochastic spike thresholding and leak values.

TrueNorth implements the LIF neuron model which supports a limited number of neuronal features (see Section 6.2) and can therefore exhibit limited spiking patterns. However, it was shown that by combining TrueNorth's neurons with carefully tuned parameters, LIF neurons could be used to express a diverse set of neuronal response patterns, including regular spiking, fast-spiking, and chattering (see Section 5.2). Details are available in [51].

Neuromorphic hardware excels in scalability. The TrueNorth core is essentially one computing block which can communicate with other TrueNorth cores, creating a mesh of distributed interconnected cores. A neuron on any core can communicate spikes to an axon on any other core in the mesh network. In a TrueNorth mesh, 4,096 cores are connected in a 2D grid, creating a routing scheme in which each core can send spikes to itself and its four neighboring cores. The encoded address is, therefore, a little more complicated than simple encoding of source's address. Each spike packet carries a dx and dy values, representing a

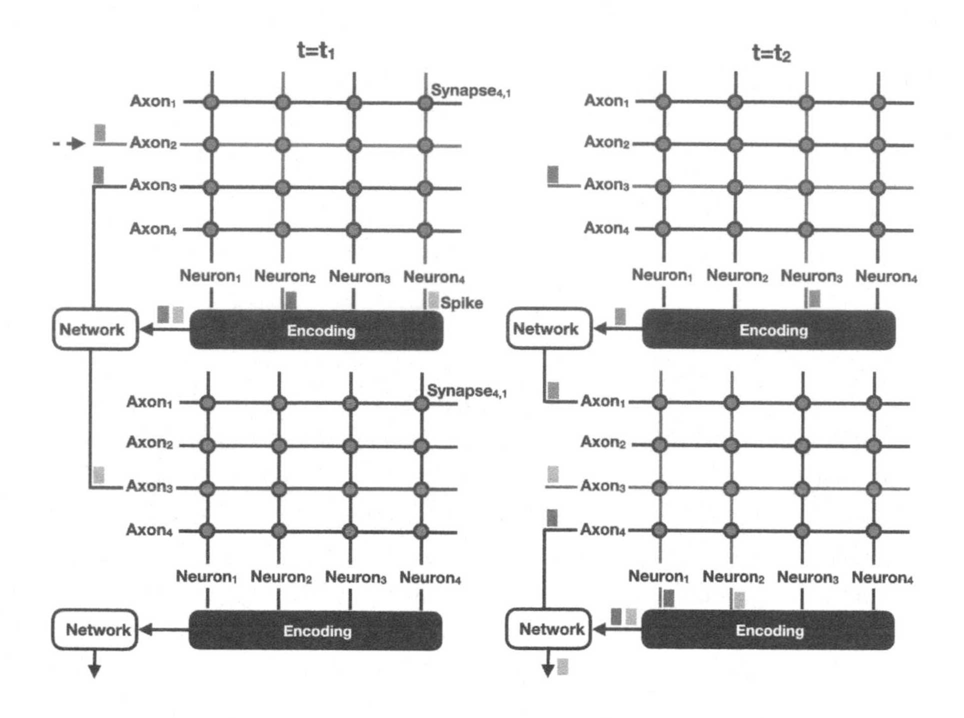

Figure 9.7 TrueNorth system schematic, showing how spikes propagate through two connected cores in two consecutive time steps. At t_1, axon 2 of one core is activated, causing neurons 2 and 4 to emit spikes. One spike is routed back to the same core at axon 3, activating neuron 2 at t_2. The second spike is routed to the second core at axon 3, activating neurons 1 and 2 at t_2. At t_2, the generated spike at the first core is routed to axon 1 of the second core. Simultaneously, two spikes are generated by the second core. One of them is routed back to the same core at axon 4 and the other is routed offcore.

relative address, i.e. the number of core-steps that the spike should travel vertically and horizontally. Moreover, each spike holds a destination axon index and several debugging flags.

A TrueNorth system with its 4,096 cores has $256 \cdot 4096 = 1,048,576 \approx 10^6$ neurons, consuming a reported power of $\approx 100 \cdot 10^{-3}$ W, and offering 10^9-scale Synaptic Operations Per Second (giga SOPS). A scaled-up TrueNorth architecture has grown to feature $16 \cdot 10^6$ neurons in the NS16e system and to $64 \cdot 10^6$ neurons in the NS16e-4 [72]. A TrueNorth Schematic is illustrated in **Figure 9.7**.

9.3 GLOSSARY

Address Event Representation: A neuromorphic inter/intra chip communication protocol in which spike-triggered events are multiplexed on a single communication bus.

Digital neuromorphic design: A digital neuromorphic system in which both computation and communication are held digitally.

Fully-dedicated design: A network connectivity scheme in which a dedicated physical wire is allocated to each neuron-synapse connection.

Hybrid neuromorphic design: A neuromorphic design in which computation is held physically with analog circuitry, and communication is held digitally (usually with AER).

Shared axon design: A network connectivity scheme in which axons are shared among neurons, enabling log_2N wires to support fully connected N neurons.

Shared synapse design: A network connectivity scheme which is realized with memory.

9.4 FURTHER READING

- **Section 9.1**
  - A comprehensive and detailed resource to learn about AER is the book: "Event-based neuromorphic systems" given in [179].
  - Early description of AER by *Kwabena Boahen* is given in [41]. Detailed description of the encoding and decoding circuits is given in [39] and [40], respectively. The arbitration process is detailed in [42].
- **Section 9.2**
  - The TrueNorth chip design was described in a series of manuscripts by IBM research. Among them are [207] which provides a basic description of the chip; [51] which provides the details of the neurons' response patterns available with

the chip; and [205] which provides the rationale beyond the chip design. More recently, IBM released a road map for the TrueNorth development given in [72].

CHAPTER 10

In-memory computing with memristors

Abstract

The memristor, "the fourth element," is a prominent candidate for neuromorphic non-CMOS hardware. Memristors have been a subject of interest for mathematicians, physicists, material scientists, electrical engineers, and neuroscientists. Memristors' physical implementation is also broad and diverse. Memristors were proposed to have a key rule in future deep learning accelerators and, due to their synapse-like plasticity and neuron-like characteristics, they can have an important impact on SNNs. In this chapter, we will discuss the basics of in-memory computing with memristors. This chapter does not aim to give a formal and rigorous description of the memristor but rather provide some intuition regarding its potential rule in neuromorphic engineering.

10.1 FROM TRANSISTORS TO MEMRISTORS

The memristor or "memory-resistor" is a fundamental electronic circuit element which provides the fourth missing piece in the classic, well-known circuit elements list which also comprises the resistor, the inductor, and the capacitor. The memristor was theorized by *Leon Chua* in the 1970s and for decades remained an object for theoretical interest. Only in 2008, did a research group in Hewlett-Packard (HP Labs) provide a working quantum-device which operates as a memristor [309]. The memristor's *"memristance"* depends on the amount of electrical charge driven through it. It features low-power consumption, memory, compatibility with CMOS technology, and synapse-like behavior. Memristors therefore, attract immense interest across the neuromorphic research community.

We will start by introducing the memristor as the fourth fundamental circuit element, as proposed by *Leon Chua* [58]. Then, we will briefly describe some key ideas for the implementation of memristors which are rooted in quantum physics. Finally, the utilization of memristors in neuromorphic hardware will be described.

Chua's description of the memristor is remarkably technical, and therefore requires broad preliminary knowledge and a significant investment of time and energy to fully comprehend. The interested reader with the appropriate background is encouraged to consult this chapter's Further Reading list.

10.2 A NEW FUNDAMENTAL CIRCUIT ELEMENT

In the classical circuit theory, the three well-known fundamental elements, which have also been extensively used throughout this book, are the resistor, capacitor, and inductor. These three circuit elements are passive, as no internal power source is required for their operation. These circuit elements relate current, voltage, charge, and flux linkage. Among these constituents, the following five pairwise relations are well known:

1. **Current**. The rate of change, or the flow, of charge q defined with:

$$i = \frac{dq}{dt} \tag{10.1}$$

2. **Resistance**. Defined by the resistor. Resistance relates current to voltage according to Ohm's Law:

$$i = \frac{v}{r} \tag{10.2}$$

3. **Capacitance**. Defined by the capacitor which stores electrical energy in an electrical field. Capacitance relates charge to voltage with:

$$C = \frac{q}{v} \tag{10.3}$$

4. **Induction**. Defined by the inductor. Induction relates current with a *flux linkage* φ, which most engineers will better recognized as its derivative, the *magnetic flux linkage* φ_B, where:

$$L = \frac{\varphi_B}{i}. \tag{10.4}$$

Briefly, when current i is driven through an inductor (or any other conductor), it generates a *magnetic field* surrounding it. When this current fluctuates (or changes), the magnetic field changes as well. A change in the magnetic field generates an opposing Electro-Motive Force (EMF), voltage across the inductor. The inductor therefore opposes changes in the current.

5. **Voltage**. EMF is described by *Faraday's law of induction*, one of the fundamental laws of electromagnetism. If we integrate the voltage over time, we derive the flux linkage φ:

$$\varphi = \int v dt \tag{10.5}$$

The relation between *flux linkage* and voltage can therefore be defined with:

$$v = \frac{d\varphi}{dt} \tag{10.6}$$

While these five relations are well known, there is a missing piece required to complete the symmetry of these relations. This missing piece was proposed by *Leon Chua* as the memristor: the device that relates flux linkage with charge (**Figure 10.1**).

Four decades later, researchers in HP labs provided a possible implementation for the memristor: the *HP memristor model*. The HP memristor is based on a layered fabrication of a platinum–titanium-oxide–platinum nano-scale device. The model was proposed in [270], in an article titled: "The missing memristor found." The finding was summarized in [281], in an article entitled: "The Fourth Element."

When an alternating voltage is applied across the memristor, its conductance follows a non-obvious curve. As the direction of voltage changes, the memristor switches its conductance state - from being less to more conductive. However, when switching the direction of the voltage, the memristor's resistance does not follow the exact resistance path it took reaching there (**Figure 10.2**).

Interestingly, as the input frequency increases, the memristor becomes more linear (**Figure 10.3**). The regular linear resistor is, therefore, a memristor at its limit frequency.

This "hysteresis" effect defines the memristor as a nonlinear resistor. The memristor's memristance depends on the history of the voltage across it — its name, a contraction of "memory resistor" reflects that property [281]. The memristor has a non-volatile memory, as its input is

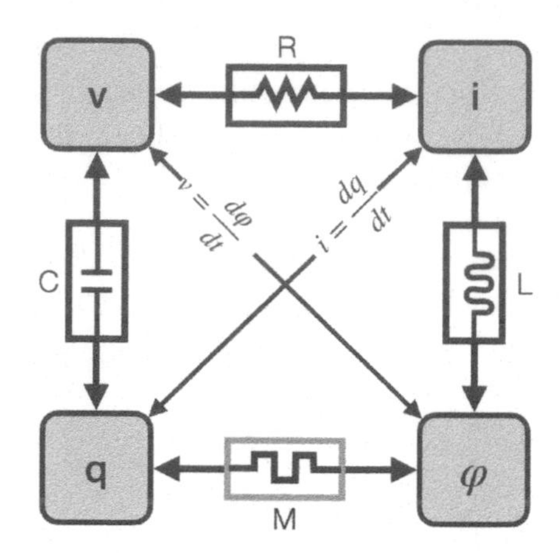

Figure 10.1 The three fundamental circuit elements are the resistor, capacitor, and the inductor. These elements relate current, voltage, charge, and flux linkage. Five pairwise connections between theses elements are well-known. The sixth missing link is the memristor which relates charge to flux linkage.

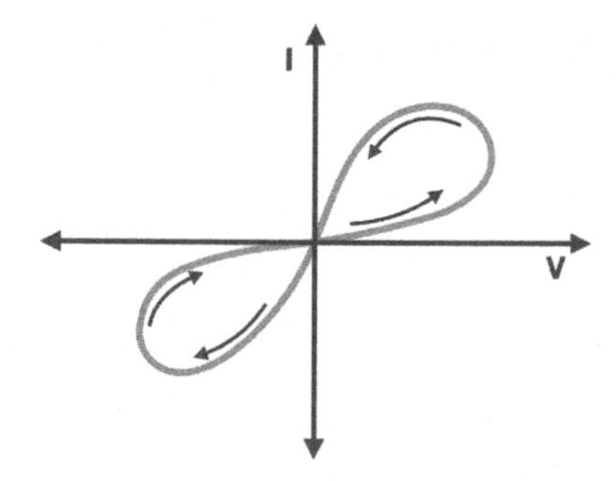

Figure 10.2 The memristor V-I curve exhibits an hysteresis effect, causing its memristance to change non-linearly with respect to the applied voltage. The memristor's memristance depends on the previous flux of charge the memristor experienced.

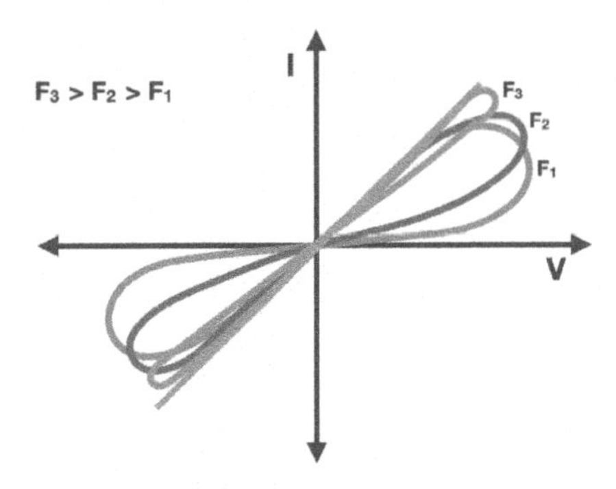

Figure 10.3 While the memristor's memristance changes in a non-linear pattern with respect to the applied voltage, as the frequency of the input voltage increases, this pattern becomes more linear.

removed, the device maintains its resistance or memristance indefinitely, changing only when an input with sufficient duration reverses its state.

Similar to resistors, memristors can be connected as networks. They follow the same resistance rules applied to regular resistors when they are connected in a series (equivalent resistance R_T is defined with: $R_T = R_1 + R_1 + ...$) or in parallel (where $1/R_T = 1/R_1 + 1/R_2...$).

Many believe that memristors will revolutionize computer memory hardware. Moreover, since two memristors can be combined to create a smaller transistor, this technology can be used to address some of Moore's law's limitations, described in Section 2.1.

In **Figure 10.4**, the way a memristor can be utilized to store 4 binary states $((00)_2, (01)_2, (10)_2, (11)_2)$, with four memristance values $(R_{00}, R_{01}, R_{10}, R_{11})$ is illustrated. Each resistance value is associated with a particular binary state. A voltage with a particular direction, amplitude, and duration has to be applied to change the memristor state.

10.3 MEMRISTORS FOR NEUROMORPHIC ENGINEERING

Memristors can be thought of as equivalent to biological synapses. Like the memristor, a synapse can modulate its weight. Weight can be modulated during learning in accordance to a learning rule or to some nonlinear function which considers the synaptic input history. Memristor-based synapses are directly implemented in hardware, allowing conductance to change during learning. Neuromorphic hardware designs can utilize

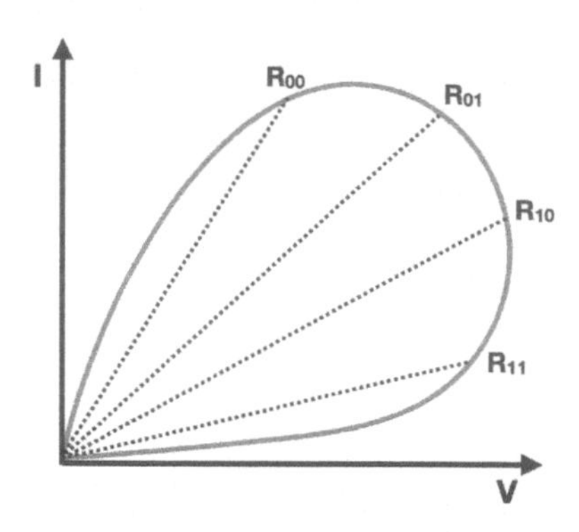

Figure 10.4 The memristor's memristance can be used to represent a particular memory state, thus enabling a memristance-memory mapping. Here, the memristance values $(R_{00}, R_{01}, R_{10}, R_{11})$ represent four binary states $((00)_2, (01)_2, (10)_2, (11)_2)$, respectively.

transistors to implement neurons and memristors to implement the required dense array of synapses [148].

A popular configuration of a neuromorphic CMOS-memristive circuit is the crossbar configuration (similar to the crossbar architecture of the TrueNorth chip (**Figure 9.6**)). It comprises a 2D array of memristor synapses, formed at the intersection of layered CMOS-based pre- and postsynaptic neurons. Neurons are connected to the crossbar with nanowires (e.g., carbon nanotubes), essentially creating a 2D array of cross wires [182]. With the crossbar architecture, every wire in the "presynaptic" neuron layer of the crossbar is directly connected to every wire in the "post-neuron" neuron layer with unique synaptic weights (**Figure 10.5**). A synaptic density of 10^{10} per cm^2 fabrication was demonstrated in [148].

To implement memristive-based synapses, memristor conductance has to be modulated by incoming spikes. However, spikes have to be modified to be *positive* to increase synapse conductance and *negative* to decrease conductance. Moreover, mV-scale spikes are not necessarily sufficient to drive conductance change. For example, the Ag-Silicon-based memristive design proposed by *Lu Wei* and colleagues had to generate -2 V and 4 V spikes to drive $\approx$ 15% conductance change. However, as the field continued to advance, new memristor designs emerged and these constraints were slowly alleviated.

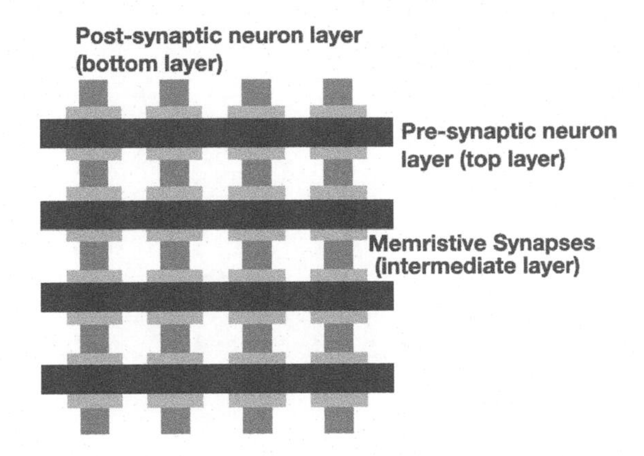

Figure 10.5 The memristor-CMOS crossbar provides an efficient fully connected neuronal architecture in which the top layered wires are connected with an intermediate memristive layer to wires in the bottom layer.

10.4 GLOSSARY

Memristor: A non-linear two-terminal fundamental circuit element which relates electric charge and magnetic flux linkage.

10.5 FURTHER READING

- A comprehensive source for data on the memristor is the book "Memristor networks," edited by *Leon Chua* and *Andrew Adamatzky* [6].
- An entering point to the world of the memristor is the article: "Memristors for the curious outsiders" by *Francesco Caravelli* and *Juan Pablo Carbajal* [50].
- An excellent review on memristor technologies and potential applications is given in [204].

IV

The Algorithms Designer's Perspective

CHAPTER 11

Introduction to neuromorphic programming

Abstract

This chapter will take the first few steps toward neuromorphic programming. We will start with emerging theoretical concepts on neuromorphic computing (e.g., complexity theory) and move on to neural codes. Key neuromorphic programming paradigms will be discussed, ranging from low-level PyNN programming which was utilized for the SpiNNaker to Corelet, developed by IBM Research to support high-level neuromorphic programming for the TrueNorth chip. Finally, we will discuss key algorithms for training SNNs toward supervised neuromorphic machine learning.

11.1 THEORY OF NEUROMORPHIC COMPUTING

11.1.1 Neuromorphic computing as Turing complete

When we think about the set of capabilities neuromorphic systems have, we usually refer to neuromorphic sensing (e.g., retina, cochlea) and neurorobotics which favor energy efficiency over performance [313] (Section 2.3). However, the most fundamental questions one can ask about neuromorphic computers are: "What can they compute? What are the computational boundaries of a neuromorphic machine? Can it compute anything computable or does it have fundamental limitations?"

Such questions are fundamental in computer science, and they usually lead very quickly to *Turing machines*. The Turing machine was defined by *Alan Turing*, back in 1936, in one of the most important works in the history of computer science theory: "On computable numbers, with an application to the Entscheidungsproblem" [287]. A Turing machine is a mathematical model of computation in which symbols can be manipulated on an infinite discretized memory tape, according to a set of rules. The Turing machine and its variants are discussed in countless books, movies, and tutorials, and they are not the focal point of this chapter (see Further Reading to learn more). Briefly, a Turing machine was shown to execute any digital algorithm. A computing system that can simulate any Turing machine is said to be *Turing-complete* or *computationally universal*. For example, von-Neumann-based ANNs were shown to be able to simulate all Turing machines using a recurrent architecture [260, 108] and can therefore be used to execute any given algorithm. The question is, therefore: Is *a neuromorphic computer Turing-complete?* The short answer is **yes**.

Continuous SNN-based dynamical systems (described in Chapter 12) were shown to be Turing complete [295, 34]. Moreover, *Wolfgang Maas* provided a general Turing complete SNN model [184] and *Aaron Russell Voelker* and *Chris Eliasmith* showed that the SPA, which was built on the foundations of NEF, can be used to construct a Turing machine [293]. However, implementing traditional Turing machines in SNNs has two fundamental problems:

1. **In memory computing architecture.** Formally, Turing machines are usually concerned about the problem: can a machine M accept input i using resources at most R? Can M decide whether i is in "language" L? In a universal Turing machine, the input, usually binary-encoded, is written on the machine's tape while the algorithm for accepting inputs is represented by M's state (**Figure 11.1**). However, in a SNN, input is not presented on a binary tape. It is encoded in the network's structure and introduced as spike trains (in-memory computing). SNN-based computing therefore requires the formulation of a new computing abstraction.

2. **Applicable realization.** From the simple mechanistic description of the Turing machine, it is not surprising to find out that a dynamical system can be used to emulate it. However, it is not clear to what extent we should be impressed by it. For example, it was shown that a particle moving in a three-dimensional space (and therefore has three DOF) could be used to build a model equivalent to a Turing machine and so be capable of universal computation

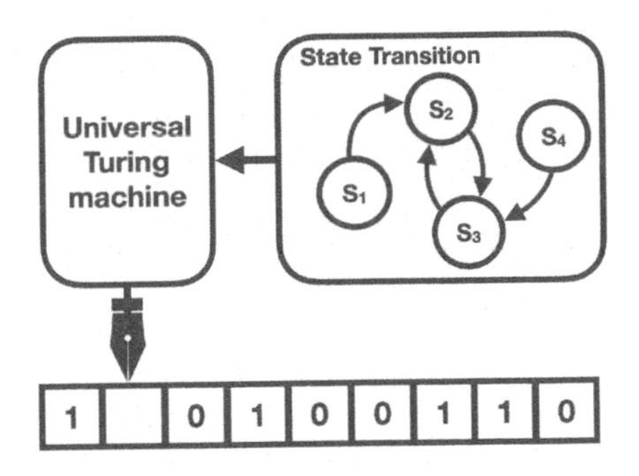

Figure 11.1 A schematic of a universal Turing machine.

[212]. Therefore, this single-particle system can be programmed to solve arbitrary problems, such as finding the millionth digit of π or determining by an exhaustive search which side, if any, has a winning strategy in chess [34]. However, it would be hard to imagine this single-particle computer being used in physical reality for more than a few consecutive computing steps. To what extent can SNN-based computing can be efficiently realized?

To generally characterize spiking neurons' computational power, we need a computation model which is anchored on a neuromorphic complexity theory. Fortunately, the way toward such complexity theory was paved by *Johan Kwisthout* and *Nils Donselaar* in their work: "On the computational power and complexity of Spiking Neural Networks" [162]. Given that a SNN S_i encodes both input i and the algorithm itself, one possible computing abstraction is this: we shall use Turing machine M to *generate* a SNN S_i which decides whether $i \in L$. However, what would be the computability of S_i? To answer that, we need a unified neuromorphic complexity theory, which will be the topic of the next section.

11.1.2 A complexity theory for neuromorphic computing

When algorithm designers try to understand how complex or how hard a problem is, they usually refer to its class of complexity. Numerous books are written on the complexity and computability of problems, and the interested reader should be able to find the one appropriate to his background and level of interest.

Briefly, in computational complexity theory, all problems which can be solved by a Turing machine using a polynomial amount of space, are in class PSPACE. We refer to a problem as belonging to *class P* if it is possible to solve (or decide) it in polynomial time, or in $O(n^k)$ where n is the size of the problem and k is a constant. Problems like Conway's Game of Life (given an initial configuration, a particular cell, and a time t, is that cell alive after t steps?) is solvable in polynomial time and is therefore in class P. Other problems which may or may not be solved in polynomial time but are verifiable in polynomial time, belong to *class NP*. The jigsaw puzzle is one example of an NP problem: we can easily verify if a solution to a puzzle is correct (go through each piece and make sure it is correctly connected to its neighbors; ($O(n)$)), but it would be much harder to find out if it is solvable. Finding out if a puzzle is solvable by an exhaustive search would take $O(4^n \cdot n!)$ operations. An easily verifiable problem necessarily has an easy solution. This is one of the most famous open questions in computer science theory: the *P versus NP problem*. Some problems in NP can be classified in the subclass *NP-complete*. A problem is in NP-complete if any problem in NP can be reduced to it in polynomial time. They form the group of the hardest problems in NP in which the Boolean SATisfiability problem (SAT) was the first member. SAT is, therefore, the *hardest* problem in NP. If we could efficiently solve SAT, we could efficiently solve every other problem in NP. Another interesting class of complexity is *class BQP*, the class of problems solvable by a quantum computer in polynomial time, with some bounding error probability. The relationships between these classes are shown in **Figure 11.2**.

These classic complexity classes take into consideration *time* and *space*. While time and space are essential resources, neuromorphic architectures consider another essential resource: *energy*. Unfortunately, energy is not captured by Turing machines as a relevant nor an essential resource for computation. Therefore, we need an extended theory of complexity which will capture energy as a vital resource. Such a theory will enable us to decide what problems can and cannot be solved efficiently on neuromorphic architectures. The development of such a theory is still in progress. One possible framework for neuromorphic systems was initially proposed in [162]. Note, that SNNs use supervised optimization methods to train neurons for pattern recognition (Section 3.1). In that sense, the training phase should be considered part of the system's required energy.

In the previous section, an abstraction for neuromorphic computing was described. A Turing machine M was used to generate a spiking neural network S_i for every input i (pre-processing). Following this model, while M should be bounded to resources R_T, which relate to

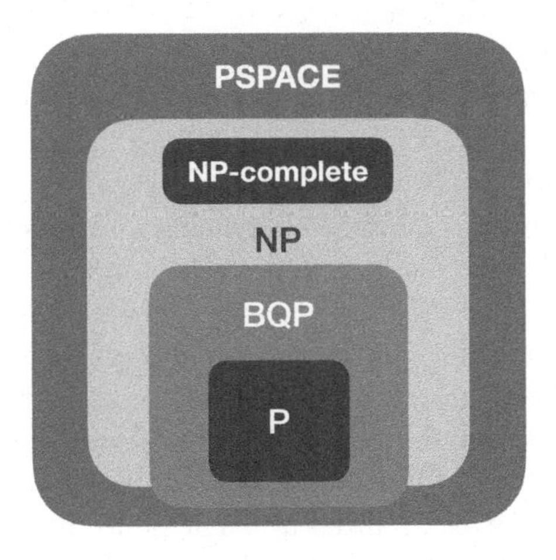

Figure 11.2 Relationships among complexity classes.

space and time, S_i is bounded to resources R_S, which relates to space, time, and energy. For R_S, *time* stands for the number of steps S_i it is using, *space* stands for the number of neurons in S_i, and *energy* for the number of generated spikes. In practice, energy is bounded by the multiplication of time and space, as each neuron can fire only once in each time step. S_i can be enhanced with energy monitoring and a timer while keeping its construction time asymptotically the same (see justification in [162]). These meters can be used to limit the allowable resources for the network before halting. This formulation might pave the way toward defining an energy-based hierarchy theorem which, similar to the time hierarchy described above, will formulate the relation between energy resources and addressable problems.

11.2 UNDERSTANDING NEUROMORPHIC PROGRAMMING

In his influential 1977 Turing Award lecture: "Can programming be liberated from the von-Neumann style?", *John Warner Backus* wonders if the art of programming can be changed - or liberated - from the "primitive word-at-a-time style... inherited from... the von-Neumann computer" [23]. The von-Neumann architecture comprises a central processing unit, memory, and an information bus connecting the two. This information bus constitutes the *memory wall* described in Section 1.2.3, and it is one of the main characteristics of the "word-at-a-time" computing paradigm.

Programming languages, such as C or JAVA, are high-level versions of the von-Neumann architecture. These programming languages use control structures (e.g., conditional *if-else* clauses and iterations) to extend machine assembly language to support irregular leaping from one place in memory to another. They use variables to represent memory registers and statements to store and retrieve data. Data entities are identified with physical addresses, creating an address-driven communication bus, in which most of the traveled information is not the data itself but rather instructions concerning where to find it. According to *John Backus*, this programming paradigm created an "intellectual bottleneck" which has "kept us tied to word at-a-time thinking instead of encouraging us to think in terms of the larger conceptual units." In many ways, neuromorphic hardware provides a different way of thinking about software and computing architecture. The neuromorphic world, however, forces us to think differently about data and algorithms.

Programming spiking neurons should address several algorithmic challenges, including:

1. Program a parallel, distributed, event-driven architecture that uses spikes and has discrete-valued synapses

2. Handle incoming data by representing it with spiking neurons

3. Adhere to hardware-specific constraints (number of neurons, axons, and synapses)

4. Optimizing implementation in terms of energy (number of participating neurons; average spike rate) and time to solution

5. Translate spiking output to a suitable format

Numerous frameworks for SNNs programming were developed. One important representative is *PyNN* [70]. PyNN provides an easy way to define populations of neurons with a high-level of abstraction. It can be used to define detailed large-scale neuronal modules, including the definition of layers, columns, and synaptic connections. It can be used to model single neurons in great detail and define templates with which ensembles of thousands of neurons can be created and connected with connectivity rules. It has a powerful compilation engine with which a PyNN-defined network can be simulated over various simulation networks, including NEURON (Section 5.3). PyNN was adopted to the SpiNNaker neuromorphic hardware using sPyNNaker - a Python library, designed to deploy PyNN-based SNNs on the board. [245]. However, it is difficult to design functional spiking circuits for applications, such as motor control or object recognition with PyNN. From the algorithm designer's perspective, choosing the right level of neuronal abstraction for network design is paramount. It is particularly important for maintaining the delicate balance between the ability to think creatively about an application and to efficiently implement it. PyNN provides an important avenue for research, as it separates software from hardware and simulation framework. However, it does not provide an easy way to define circuits with higher functional roles.

From the algorithm designer's perspective, (important) matters, such as optimization and hardware limitations should ideally be abstracted away. Frameworks for neuromorphic programming should provide a hardware-tailored abstraction for functional neural circuits. For example, each Loihi core has a programmable learning engine that can modulate synaptic state variables over time as a function of its spike activity to provide supervised learning [69] (Learning SNNs will be further discussed in Section 13.1). Intel provides a Python-based Application Programming Interface (API) with which functional learning circuits can be defined in few code lines, particularly to build, train, and use SNNs. It was demonstrated with the classification of handwritten digits from the MNIST database [176]. Such a high-level of abstraction allows for smooth, streamlined programming. For example, MNIST classification on Intel's Loihi looks like this:

1. Create an input layer for the network:
   input = snn.createStimulusNodeGroup(...).

2. Create a SNN as an output layer:
   output = snn.createNeuronGroup(...).

3. Define a learning rule:
   rule = STDPpotentiationRule(...).

4. Connect the network:
   snn.fullyConnect(input, output, rule),

5. Prepare training data:
   LineScanStimulus(training_images, input),
   and apply it to the network: *snn.addExternalStimulus(...).*

6. Set a runtime controller for training:
   snn.addRuntimeController(...).

7. Finally, simulate: *simulate(...))*,
   and retrieve resolved weights: *getStateValues(...)*.

Such a high-level description allows the algorithm developer to concentrate on his work's creative parts, such as network design, data curation, and learning rules articulation instead of on low-level technical details, such as spiking rates and time constants. While robust for certain applications, this API is tailored to specific hardware, particularly addressing its relative strength - real-time learning.

Back in 2011, IBM designed the TrueNorth chip, described extensively above. Along with the development of the chip itself, IBM developed the Neural Programming Model (NPM), aiming at providing high-level programming for the TrueNorth. As the reader might remember from Section 9.2, each TrueNorth core has at least 2^{256^2} synapse configurations, 3^{256} axon type configurations, and leakage and threshold values for each neuron. Finding the right configuration in this enormous state space while aiming at realizing one desired function is an immense challenge. However, careful mapping of spiking neurons to functions was shown to produce productive behaviors. For example, NPM was shown to control a robot driving in a virtual environment, play the classic game of pong, and recognize handwritten digits [21]. The robot control was implemented using neurons as "lane detectors," representing the road's position by having neuron spikes when the road is within its receptive field. These neurons are excitatory at the center, inhibitory at the edges, and silent elsewhere, forming a filter which fires maximally when the road is centered. Spikes are integrated to produce two values, one for each wheel: when the road is to the left, the left wheel receives higher driving power, causing the robot to turn right (**Figure 11.3**).

To enhance NPM, IBM came out with an entirely new Eco-system, comprised of a programming language, named *Corelet* [11], a Corelet library [90] and a simulation framework named *Compass* [233]. This Eco-system was designed to handle the challenges listed above. Corelet abstracts away the underlying hardware architecture. A Corelet is a virtual object which encapsulates a TrueNorth core, hiding the underlying network details and connectivity scheme, only exposing input and output ports. Each Corelet is optimized to carry out one specific, relatively simple, function. Corelet optimization can be carried on off-chip (in a regular computer), using different optimization protocols (e.g., backpropagation). IBM designed some commonly used functions as Corelets, such as filters, averaging, and conversion modules (e.g., rate to binary) [90].

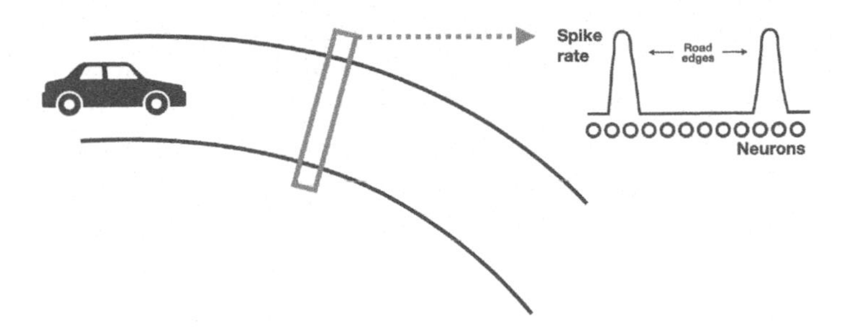

Figure 11.3 Controlling a virtual car with spiking neurons. Example from [21].

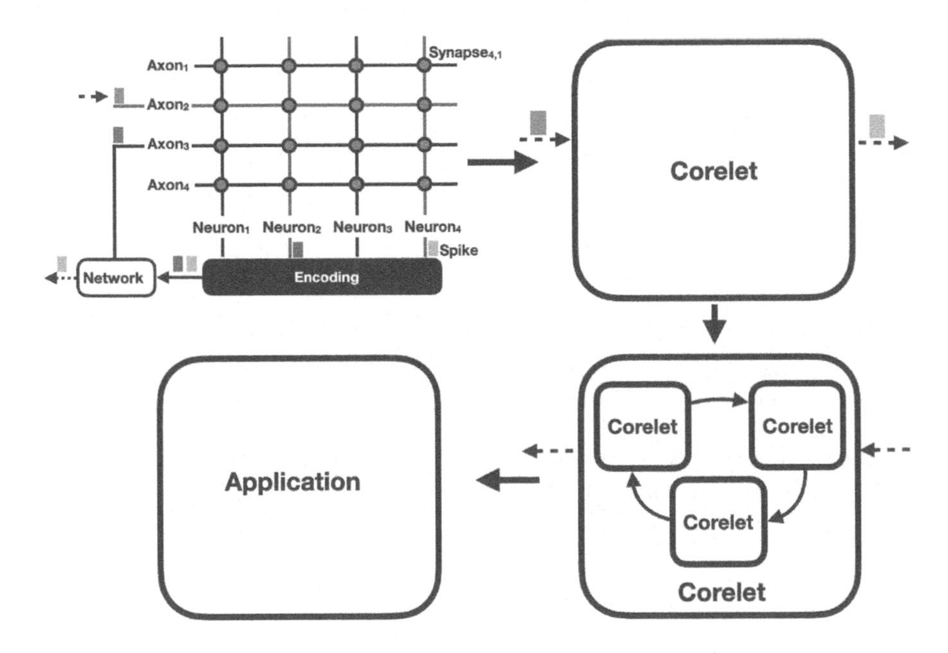

Figure 11.4 An example of Corelet composability.

The same Corelet can be used in different applications; different Corelets can be concatenated together like LEGO bricks in a great variety of ways. For example, a convolution network Corelet can be comprised of a set of convolution filters and averaging Corelets, where each convolution filter comprises a set of filter Corelets. Detailed implementation is given in [90]. The main point of discussion here is the composability of abstract entities (**Figure 11.4**). Corelets hide hardware constraints and allow algorithm designers to be focused on the application at hand. Following this Object-Oriented (OO) state of mind, Corelets can be inherited (for reuse and adaptation) and used polymorphically. Adopting the OO Programming (OOP) paradigm, which most programmers are familiar with, improve the design, structure, modularity, correctness, consistency, compactness, and reusability of neuromorphic programming.

Corelet was tailored-designed for the TrueNorth. However, a key development stage in most programming paradigms is the development of another abstraction layer: a layer that hides hardware's details such that it will no longer be a concern. In such a framework, a high-level description of an algorithm will compile on any neuromorphic board. Ideally, the same code could be executed on the TrueNorth, the SpiNNaker, the Loihi, and even on the NeuroGrid, with little to no modifications. Such a framework would have to be general enough to provide data representation and modification with spikes. This mission was taken by the NEF, which will be the topic of discussion in the following chapter.

11.3 GLOSSARY

Computational complexity theory: A theoretical framework for classifying and relating computational problems according to their resource usage.

Turing machine: A mathematical formulation of an abstract machine which manipulates symbols on a strip of tape according to a set of rules.

11.4 FURTHER READING

- **Section 11.1**
  - A well-recognized introductory book about the theory of complexity and computability is: "Introduction to the Theory of Computation" by *Michael Sipser* [262].
  - Learn more about the computability and complexity of SNNs in [162].

CHAPTER 12

The Neural Engineering Framework (NEF)

Abstract

One of the most significant steps in software design was the development of the compiler. The compiler translates and optimizes high-level instructions to machine code. It separates code from the underlying hardware such that high-level descriptions can be compiled to work on various hardware architectures, ideally, with no changes to the code itself. In the neuromorphic world, a neuro-compiler would translate a high-level description of a SNN into low-level neuron specifications. One framework which allows such compilation is Nengo, powered by the NEF. In this chapter, we will learn about the basics of the NEF and see how it can come to life with Nengo. This chapter's didactic approach was inspired and modified (with permission) from the course notes for SYDE 556: Simulating Neurobiological Systems, offered by Chris Eliasmith at the University of Waterloo.

12.1 THE FUNDAMENTAL PRINCIPLES OF NEF

The NEF throws out neurons' anatomy and physiology. It identifies the neuron's computational essence and encapsulates it within a point-process. NEF provides a computational framework that allows efficient use of many such processes to approximate complex systems' dynamics. Particularly, NEF provides three fundamental principles which guide the assembly of large scale neuronal models. In contrast to traditional bottom-up modeling, NEF takes the top-bottom approach, in which high-level descriptions constraint the network's low-level building blocks (see Section 1.2.2). NEF has been used to demonstrate increasingly complicated dynamical systems, including path integration, working memory, list memory, inductive reasoning, motor control, and decision making [87]. NEF is described in length in [84] and succinctly reviewed in [267].

The NEF proposes three principles which guide the design of large-scale neuronal models:

1. **Representation.** A population (or ensemble) of neurons can distributively represent signals through nonlinear - spike-based - data encoding and decoding. In an encoding phase, mathematical constructs are represented by sets of spike trains and, in the decoding phase, these spike trains are filtered and weighted summed up. Weights can be optimized to reconstruct the represented signal or to achieve some desired transformation of that signal.

2. **Transformation.** Linear and nonlinear functions of encoded signals can be implemented though weighted synaptic connections between ensembles of neurons.

3. **Dynamics.** Connecting ensembles through weighted synapses can be used to create feedback (recurrent) connections, thus creating a dynamical system. Such a dynamical system can sustain neural activity, providing means for high-level functionalities, such as memory.

Nengo is a Python-based software library which provides an API to NEF, thus allowing the simulation of large-scale neuronal models. With Nengo, ensembles of neurons and connections among these ensembles can be trivially defined. Nengo provides a set of functions that realize encoding, decoding, and transformation in the back-end, allowing the developer to concentrate on the neuronal model itself. In the following sections, we will dive into the depth of NEF and demonstrate pieces of it with Nengo. Code is available on the book website.

12.1.1 Representation

12.1.1.1 Encoding

Let x be the value which we want to represent with an ensemble of spiking neurons. Representation a of x would take the form of:

$$a = f(x) \tag{12.1}$$

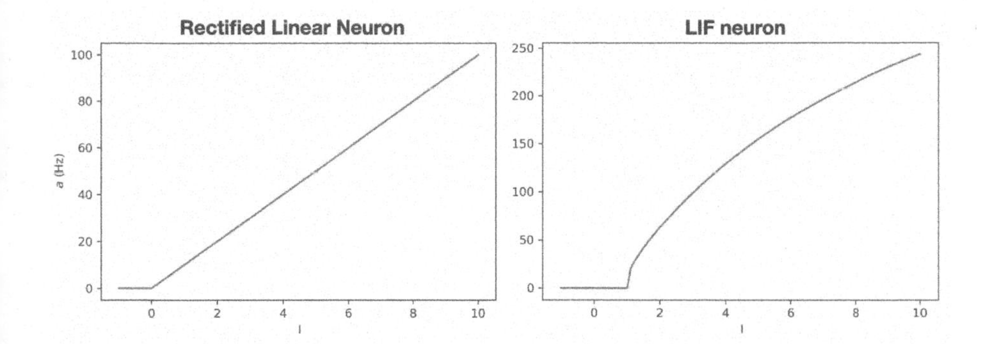

Figure 12.1 Linear rectified (left) and LIF neuron models (right). The linear rectified model was defined with gain= 10 and $b = 0$ (bias). The LIF neuron was defined with $\tau = 0.02$ and $t_{ref} = 0.002$.

A **neural** representation would be in the form of:

$$a = G(J(x)) \tag{12.2}$$

where G is a neuron model, and J is the stimulus, driven into the neuron model.

One simple way of defining a is using a *rectified linear neuron* in which neuron activity initiates at $J = 0$, linearly increasing afterwards (**Figure 12.1, left**). Alternatively, a biologically plausible neuron model is LIF, for which we can use Eq. 5.23. In NEF, LIF neurons are normalized to feature a resting state at 0 and a firing threshold at 1 (arbitrary units) (**Figure 12.1, right**).
LIF neuron model response to a sinusoidal voltage is demonstrated in **Figure 12.2**.

In NEF, a stimulus is represented distributively, as the same value is driven to multiple neurons:

$$a_i = G_i(J(x)) \tag{12.3}$$

One simple current model $J(x)$ is:

$$J(x) = \alpha x + J^{bias} \tag{12.4}$$

where α is a gain factor and J^{bias} is a current bias term. In Section 1.1, the concept of each neuron having its own preferred stimulus was described. Formally, response to a preferred stimulus can be expressed with:

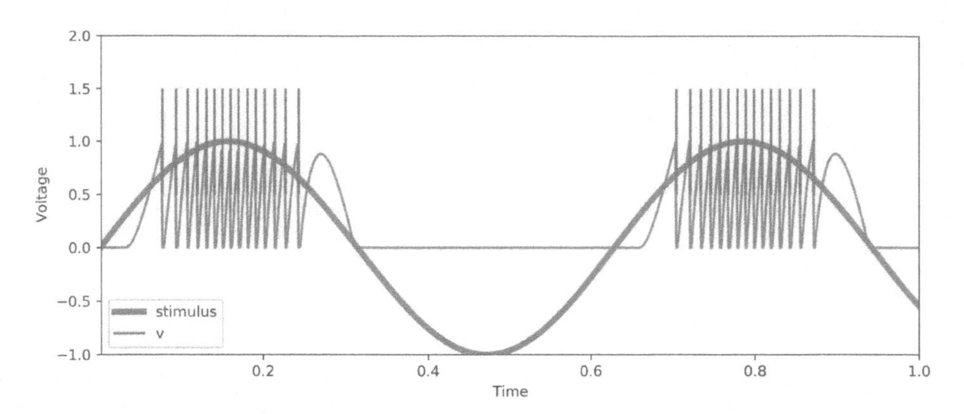

Figure 12.2 LIF neuron model (specified in Figure 12.1) response to a sinusoidal input. Spikes are shown in blue, neuron voltage in green, and input stimulus in red.

$$J = \alpha[sim(x, x_{pref})] + J^{bias} \tag{12.5}$$

where sim is some similarity function, maximized when x and x_{ref} (the preferred stimulus) are similar. x_{ref} can be defined using an encoder e, a vector representing the value to which a neuron will respond with the highest frequency of spikes. A similarity function can be defined using the dot product between x and x_{ref}, resulting in:

$$a_i = G_i(\alpha_i(x \cdot ei) + J^{bias}) \tag{12.6}$$

A useful way of representing a neuron's response to varying inputs is by using a response or a tuning curve. In NEF, a tuning curve is a fundamental concept, defined using an *intercept*, the value for which the neuron starts to produce spikes at a high rate, a *maximal firing rate* and an *encoder*. An ensemble of neurons is comprised of many neurons, each featuring a different tuning curve. Tuning curves of two neurons, featuring the same intercept (0.5) and two opposing encoders (−1 and 1) are shown in **Figure 12.3**. The response of these two neurons to a sinusoidal input is demonstrated in **Figure 12.4**. Similarly, tuning curves for 50 different neurons with uniformly distributed maximal spiking rates and randomized intercepts are shown in (**Figure 12.5**). The response of these 50 neurons to a sinusoidal input is demonstrated in **Figure 12.6**.

Eq. 12.6 proposes a powerful way to represent a stimulus, offering a robust mathematical description of data *encoding*. However, it has

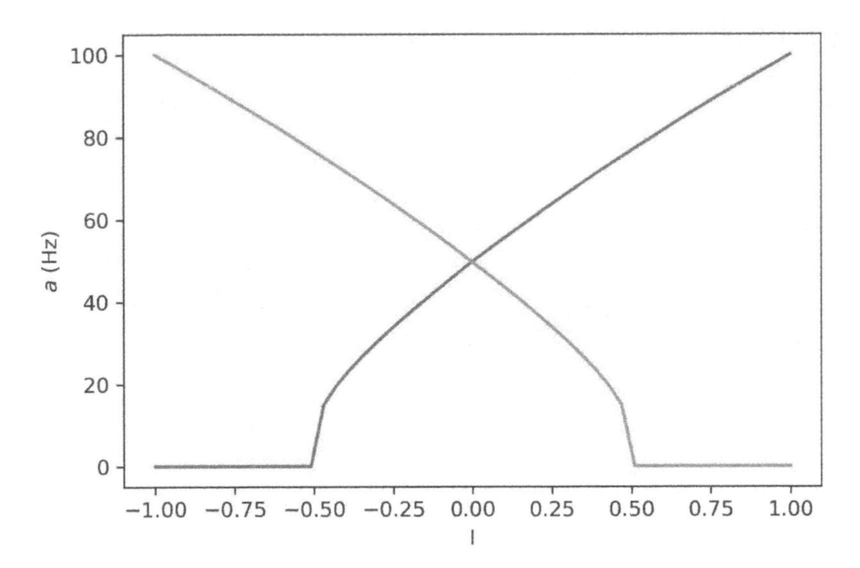

Figure 12.3 Two LIF neurons with the same intercept (0.5) and two opposing encoders (−1 (orange) and 1 (blue)).

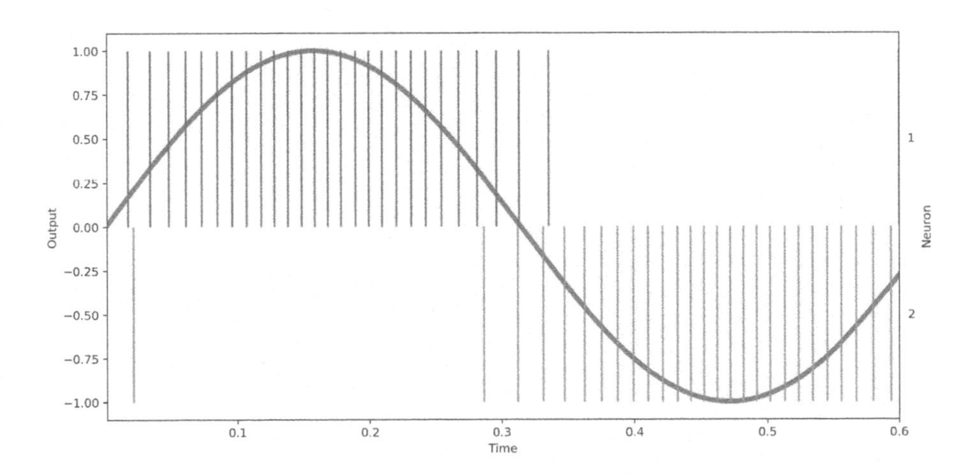

Figure 12.4 Two LIF neurons' (specified in **Figure 12.3**) response to a sinusoidal input. Spikes for the positively encoded neuron are shown in blue and spikes for the negatively encoded neuron are shown in orange.

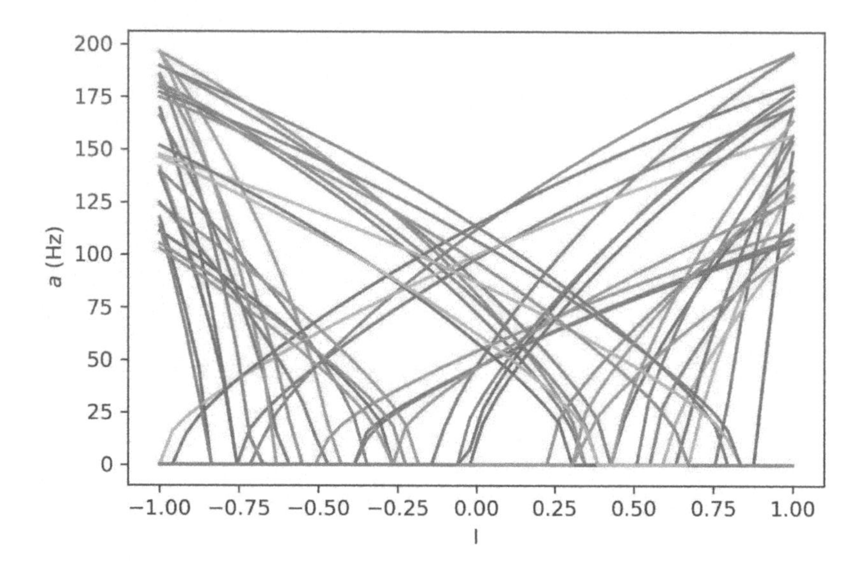

Figure 12.5 50 LIF neurons with uniformly distributed maximal spiking rates and randomized intercepts.

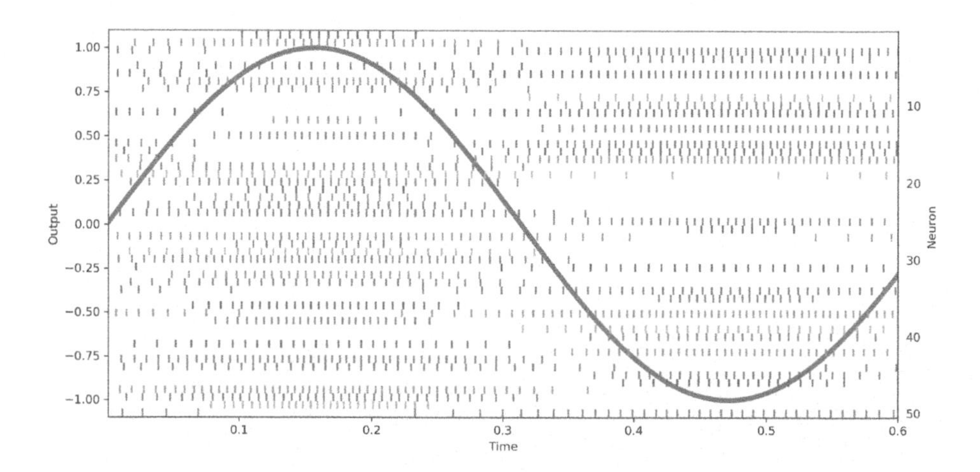

Figure 12.6 50 LIF neurons' (specified in **Figure 12.5**) response to a sinusoidal input. Spikes of each neurons are shown in a different color.

several weaknesses. Among them are the neurons being "point" processes. In contrast to the models demonstrated in Chapter 7, NEF-neurons have no morphology nor compartments. Furthermore, their neuronal characteristics (τ, u_{th}) are constants, preventing spiking adaptation.

12.1.1.2 Decoding

Data representation is comprised of both *encoding* and *decoding.* In the previous section, we were able to encode a stimulus with non-linear entities - spiking neurons. Now, we will explore how this non-linear encoding scheme can be linearly decoded.

For linear decoding, we define $\hat{x}$ as the approximated encoded x:

$$\hat{x} = \sum_{i}^{n} a_i d_i \tag{12.7}$$

where n is the number of neurons, a_i is the activity - or spiking rate - of neuron i, and d_i is the decoder for neuron i. Decoders are optimized, such that $\hat{x}$ will approximate x accurately as possible. Particularly, the decoders are calculated by minimizing the average error E over x, defined using the mean squared error: $E = \frac{1}{2}\int_{-1}^{1}(x - \hat{x})^2 \, dx$. Following Eq. 12.7 we derive:

$$E = \frac{1}{2}\int_{-1}^{1}(x - \sum_{i}^{n} a_i d_i)^2 \, dx \tag{12.8}$$

Differentiating E with respect to each decoder d_i gives:

$$\frac{\delta E}{\delta d_i} = \frac{1}{2}\int_{-1}^{1} 2(x - \sum_{j}^{n} a_j d_j)(-a_i) \, dx \tag{12.9}$$

where j represents the indices of all neurons excluding neuron i.

Eq. 12.9 can be reformulated as:

$$\frac{\delta E}{\delta d_i} = -\int_{-1}^{1} a_i x \, dx + \int_{-1}^{1} \sum_{j}^{n} a_j d_j a_i \, dx \tag{12.10}$$

Minimal error is achieved when $\frac{\delta E}{\delta d_i} = 0$, resulting in

$$\int_{-1}^{1} a_i x \, dx = \int_{-1}^{1} \sum_{j}^{n} a_j d_j a_i \, dx \tag{12.11}$$

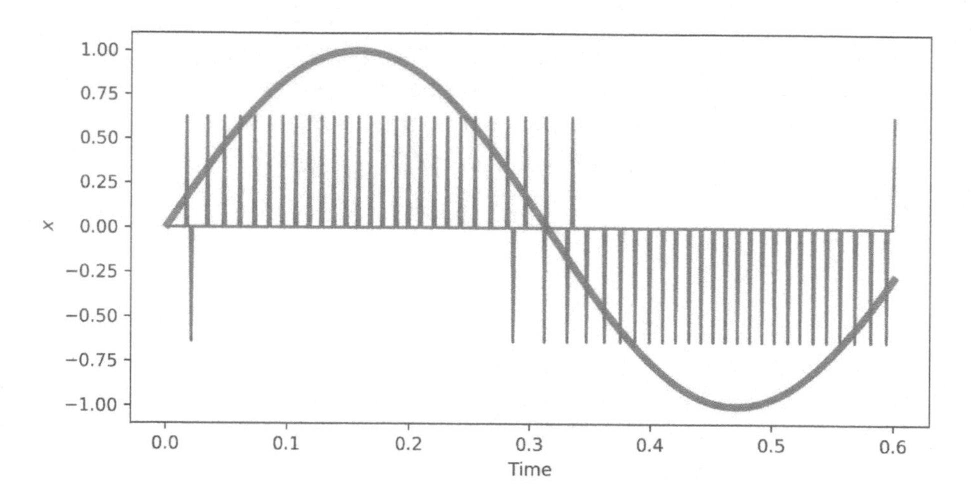

Figure 12.7 Two LIF-based stimulus decoding. Encoding is shown in **Figure 12.4**. Stimulus is shown in red and the decoded value is shown in blue.

Eq. 12.11 can be reformulated as:

$$\int_{-1}^{1} a_i x \, dx = \sum_{j}^{n} \left(\int_{-1}^{1} a_j a_i \, dx \right) d_j \tag{12.12}$$

Eq. 12.12 is a system of n equations, with n variables:

$$\Upsilon = \Gamma d \tag{12.13}$$

where $\Upsilon_i = \frac{1}{2} \int_{-1}^{1} a_i x \, dx$, and $\Gamma_{ij} = \frac{1}{2} \int_{-1}^{1} a_j a_i \, dx$.

From Eq. 12.13, the decoders can be calculated using:

$$\begin{aligned} d &= \Gamma^{-1} \Upsilon \\ d_i &= \sum_j \Gamma_{ij}^{-1} \Upsilon_j \end{aligned} \tag{12.14}$$

Let's explore the utilization of Eq. 12.14 for decoding. Optimizing Eq. 12.14 to decode the encoded stimulus shown in **Figure 12.4** will result in the reconstructed stimulus shown in **Figure 12.7**. The result, so it seems, is not too impressive.

Perhaps it is a matter of the number of the encoding neurons. Let's consider decoding with the 50 neurons showed in **Figure 12.6**. The resulted reconstructed stimulus is shown in **Figure 12.8**. While the

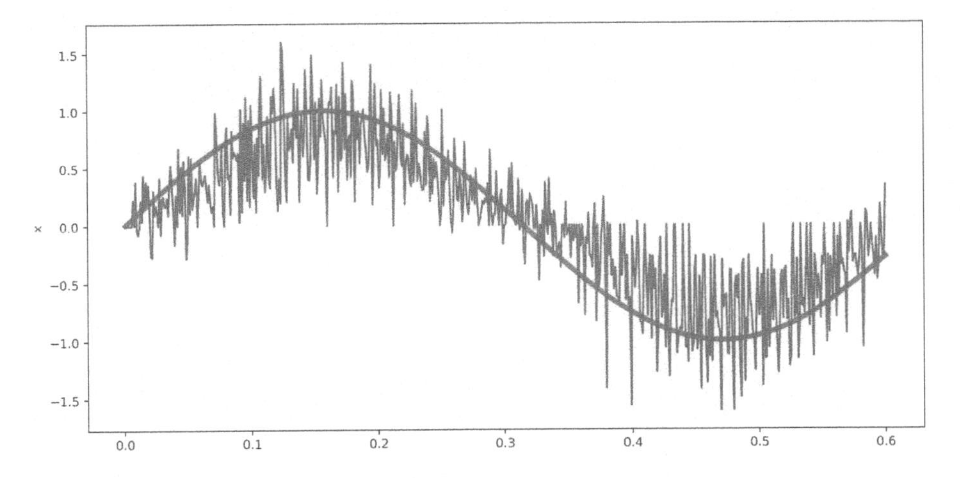

Figure 12.8 50 LIF-based stimulus decoding. Encoding is shown in **Figure 12.6**. Stimulus is shown in red and the decoded value is shown in blue.

decoded curve follows the encoded stimulus, it is not an accurate representation. The reason for this noisy representation is that in a given time window, only a small number of neurons fire spikes. As this window gets smaller, this phenomenon gets more pronounced. In some time windows, no neuron fires spikes, resulting in $\hat{x} = 0$.

To resolve this issue, we need each spike to continuously contribute to the decoding in some explicitly defined window of time. Particularly, we need to define *temporal filtering*. Temporal filtering can be mathematically formulated as convolution. With convolution, a function h modifies another function f, producing a third $f * h$ function. It is an integral transform, defined as the integral of the product of f and h, where one of them is reversed and shifted. Convolution is a weighted average of $f(\tau)$ at i where the weighting is given by $g(-\tau)$ shifted by i:

$$f * h(i) = \int_{-\infty}^{\infty} f(\tau)g(i-\tau)d\tau \tag{12.15}$$

With temporal filtering, linear decoding, denoted in Eq. 12.7, becomes:

$$\hat{x}(t) = \sum_{i}^{n} a_i(t) * h(t)d_i \tag{12.16}$$

where $*$ is the convolution operator.

Choosing $h(t)$ is a critical factor for accurate decoding. A biological plausible choice for $h(t)$ is an exponentially decaying filter, defined with:

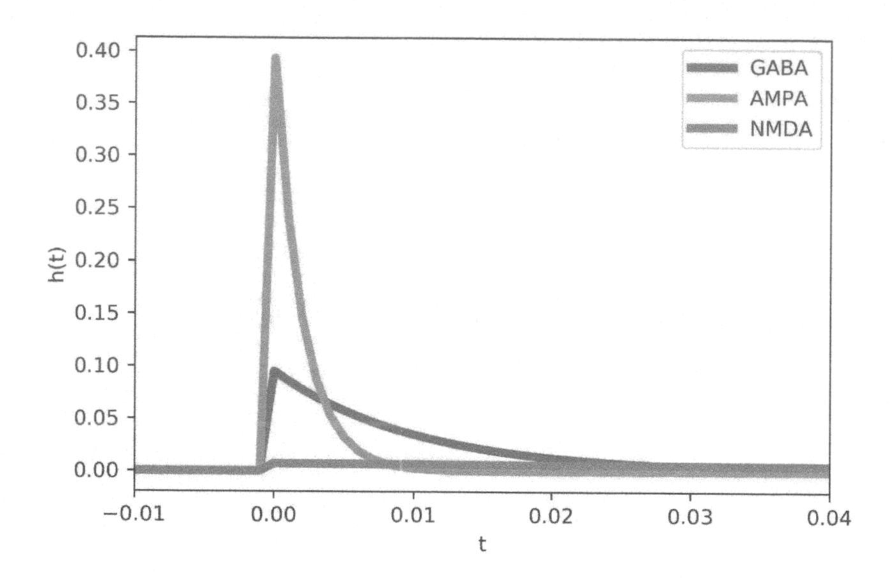

Figure 12.9 Exponentially decaying filters where $\delta t = 0.001$ and $\tau = 2, 10, 145$ ms corresponding to the AMPA, GABA, and NMDA neurotransmitters, respectively.

$$h(t) = \begin{cases} e^{-t/\tau} & \text{if } t > 0 \\ 0 & \text{otherwise} \end{cases} \tag{12.17}$$

In BNNs, τ depends on the neurotransmitter at the synapse. For example, GABA has a $\tau = 10.41 \pm 6.16$ ms, AMPA has a $\tau = 2.2 \pm 0.2$ ms and NMDA has a $\tau = 146 \pm 9.1$ ms (**Figure 12.9**).
Convolving the encoding spikes of the two neurons showed in **Figure 12.4** would result in a the decoded values shown in **Figure 12.10** (compare to **Figure 12.7**). An even more accurate decoding is achieved with a higher number of neurons. Decoding the convoluted spikes shown in **Figure 12.6** would produce an smoother reconstruction of the stimulus, as it is shown in **Figure 12.11** (compare with **Figure 12.8**).

Our neuromorphic representation seems to work well; however, as anticipated, it incorporates a slight delay and it is not a perfect match with the encoded stimulus. This is the result of a *distortion error*, induced by the decoders themselves. This *static distortion* is proportional to the number of encoding neurons n:

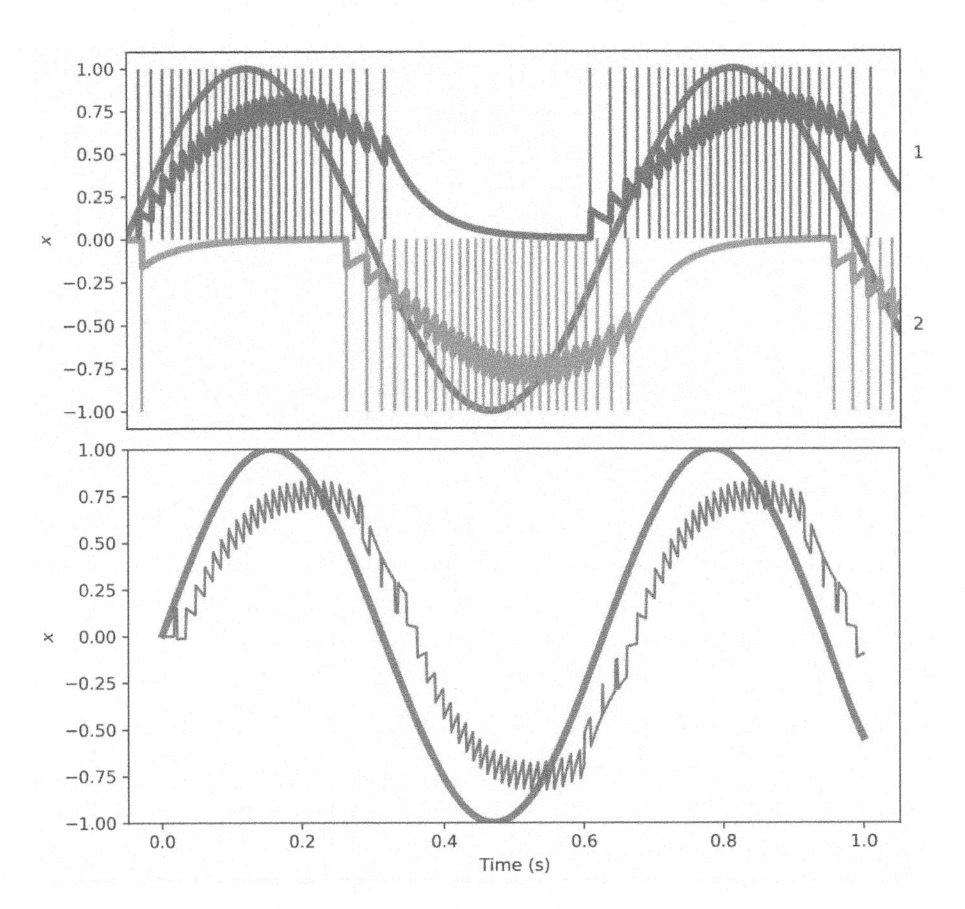

Figure 12.10 Two convolved LIF-based stimulus decodings. Positively and negatively encoded neurons are colored in blue and orange, respectively (both spikes and convolved traces are shown). An exponentially decaying filter was used with $\delta t = 0.001$ and $\tau = 50$ ms. A weighted summation of the convolved spikes is shown in the bottom figure.

$$E_{static} \propto \frac{1}{n^2} \tag{12.18}$$

As we increase the number of neurons, representation error is reduced. However, there is much more to it. The selection of the encoders and the neurons' tuning curves distribution have a drastic effect on the representation, especially in higher dimensions.

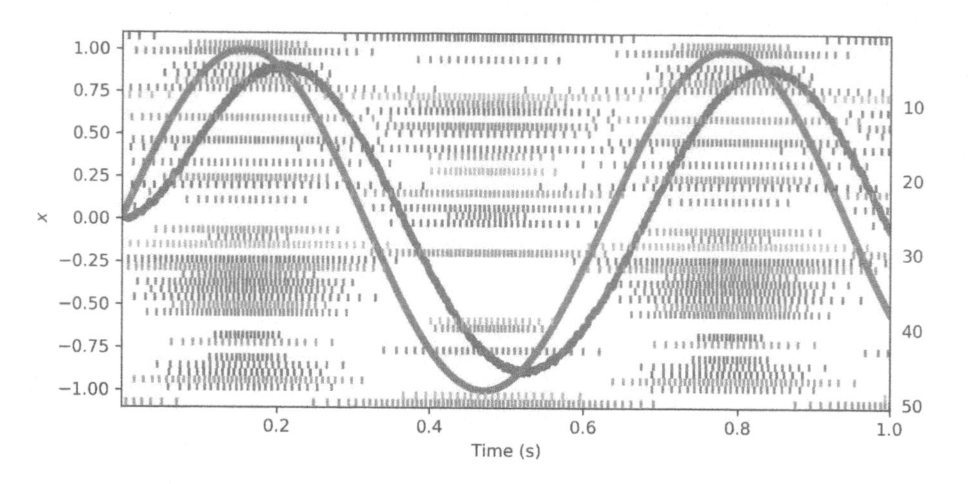

Figure 12.11 50 convolved LIF-based stimulus decoding. Spikes of each neuron are color-coded. An exponentially decaying filter was used with $\delta t = 0.001$ and $\tau = 50$ ms. A weighted summation of the convolved spikes is shown in blue and the encoded stimulus in red.

12.1.1.3 Decoder analysis

In regard to neuromorphic representation, one of the most fundamental questions is, What function $f(x)$ can we represent with spiking neurons? We will start by trying to represent a linear function $f(x) = x$. Representation with using 50 neurons with randomly distributed tunings shows accurate results with a Root Mean Square Error (RMSE) of 0.0106 (**Figure 12.12**). What would happen if we change the distribution of the tuning curves? Representation with 50 neurons with uniformly distributed intercepts across the range [0, 0.3] is shown in **Figure 12.13**. Now, the approximation has a RMSE of 0.05561. Setting all intercepts to 0.2, results in a poorer approximation with a RMSE of 0.1143 (**Figure 12.14**). We can conclude that neurons' tuning is essential for accurate representation.

To profoundly understand the limitations and power of NEF-based representation, we will investigate its underlying mathematical properties. We know from linear algebra that the standard Cartesian basis is comprised of orthonormal unit vectors which span that linear space. A 2D space, for example, is span by unit vectors $\vec{i}$ and $\vec{j}$ for which $\vec{i} \cdot \vec{j} = \langle\vec{i}\vec{j}\rangle = 0$ (they are orthogonal to each other). Considering a vector $x_{ij} = [x_1, x_2] = x_1 i + x_2 j$ in the standard Cartesian basis which we want

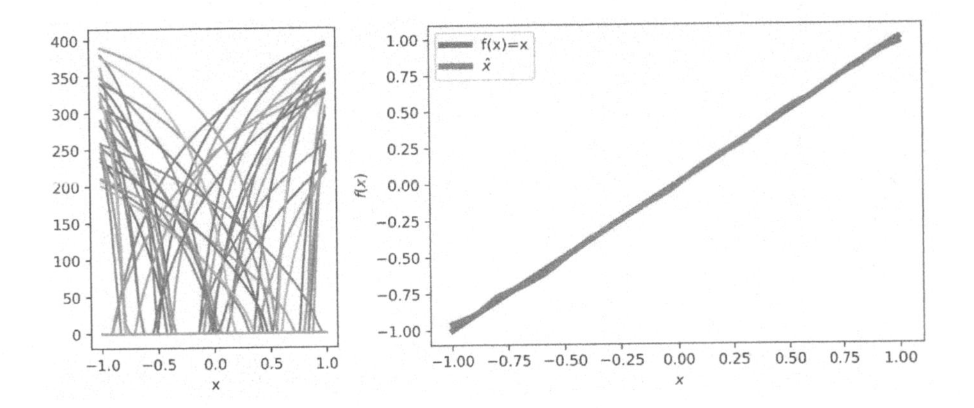

Figure 12.12 Representation of $f(x) = x$ (right) using 50 neurons with randomly distributed intercepts (left).

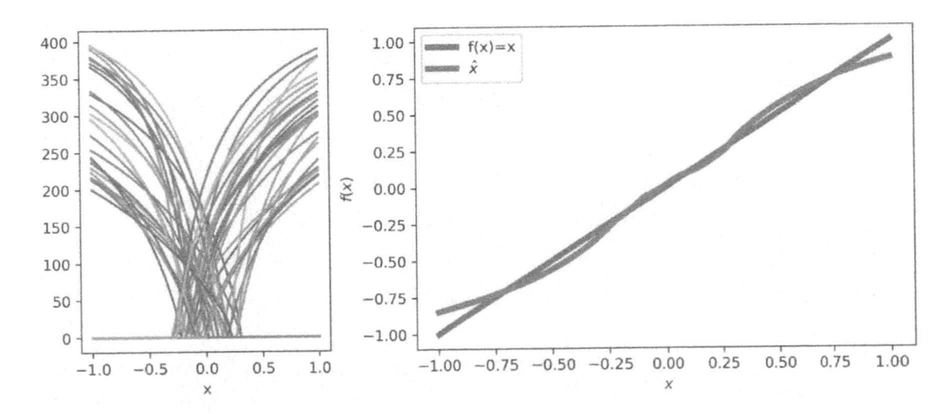

Figure 12.13 Representation of $f(x) = x$ (right) using 50 neurons with uniformly distributed intercepts at the range $[-0.3, 0.3]$ (left).

to represent in a different orthonormal basis d_1, d_2 (**Figure 12.15, left**). Projection of x onto the d_1, d_2 basis will take the form of:

$$x_{d_1 d_2} = [\langle \vec{x}\vec{d_1} \rangle \vec{d_1}, \langle \vec{x}\vec{d_2} \rangle \vec{d_1}] \tag{12.19}$$

or:

$$\begin{aligned} x_{d_1 d_2} &= [a_1 d_1, a_2 d_2] \\ a_i &= \langle x d_i \rangle \end{aligned} \tag{12.20}$$

Coefficients a_i can be considered as if they are representing or encoding x. Decoding will take the form of:

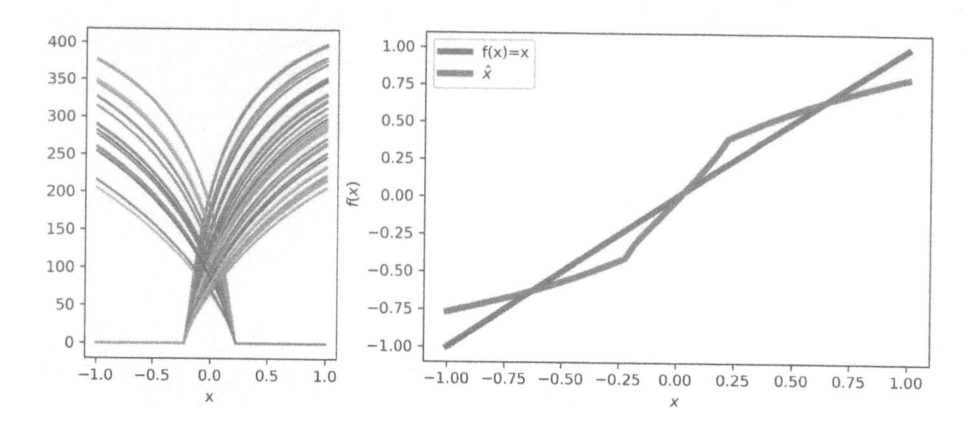

Figure 12.14 Representation of $f(x) = x$ (right) using 50 neurons with intercepts sets to -0.2 (left).

$$x = \sum_i a_i d_i \tag{12.21}$$

Why? as we substitute the encoding (Eq. 12.20) into the decoding (Eq. 12.21), we recover x (check it!). In this case, however, d_1 and d_2 were orthogonal to each other. If we relaxed this constraint, such that these vectors were not necessarily independent, we might end up with *over-complete basis*. With an over-complete representation, the number of the basis vectors is larger than the dimensionality of the input and it is, therefore, considered less efficient. For example, a point $[x, y]$ in 2D space can be represented by three equally spaced vectors $\vec{d_1}, \vec{d_2}, \vec{d_3}$ as it is shown in **Figure 12.15, right**. In noisy physical systems, however, the inherent redundancy in an over-complete basis can be proved invaluable for efficient error correction [171].

Similar to the way *basis vectors* define a vector space, *basis functions* define a function space. For example, with *Fourier analysis*, functions may be decomposed or approximated by sums of trigonometric functions (e.g., sines and cosines). We can therefore have over-complete basis functions which span some functional representation space. In NEF, neurons' tuning curves are over-complete basis functions for the space they represent. The functional space these neurons span is the space they can compute and it is therefore of key interest here. Because neural responses represent an over-complete basis, they are not-independent. Therefore, only a small part of the subspace of possible neural activity

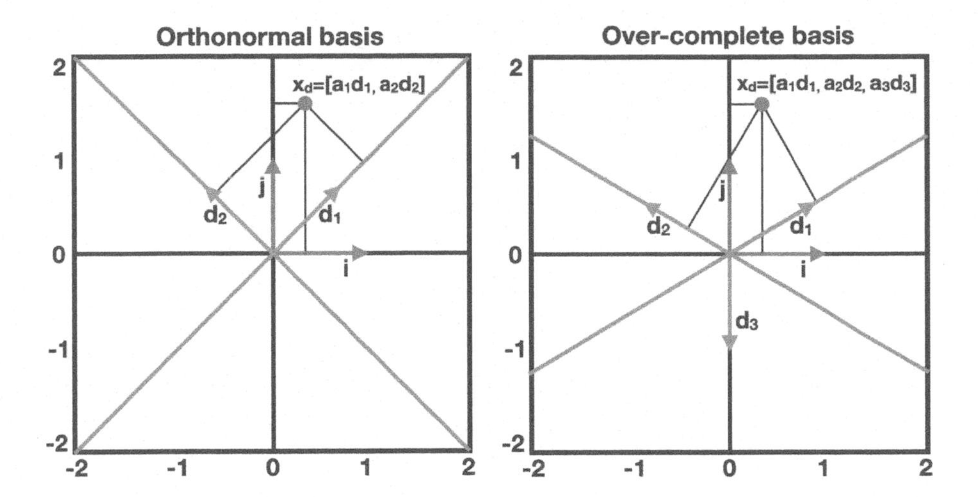

Figure 12.15 Orthogonal (left) and over-complete (right) basis in which a 2D point is represented.

values is reached by that population. Eq. 12.13 specifies the relation between the decoders and the error they induce. Particularly, Γ specifies the correlations between neurons, providing a measure of dependability needed to determine that subspace. To retrieve this information, we need to compute the inverse of the Γ matrix (needed for the decoders optimization (Eq. 12.14)). However, since neurons have similar tuning curves, Γ is likely to be singular (and it is therefore not reversible).

We can use Singular Value Decomposition (SVD) to decompose and approximately invert Γ, such that given Υ, the most appropriate decoders (in terms of minimal least squared error) are derived. The exact formulation of SVD decomposition is outside of scope for this book however, it is present in most advanced linear algebra books. In the context of this discussion, SVD returns matrices U and S. U can be thought of as a rotation matrix which aligns a coordinate system's axis according to the most considerable variance in the encoding of x. U aligns the coordinate system's first axis along the dimension with the greatest variance, the second axis along the dimension with the second greatest variance, etc. Therefore, they are sorted in the order of "importance," or independence, measured by S. S holds the error induced into the reconstruction of $Gamma$, when U is discarded. It accounts for the level of independent information projected onto that vector. By rotating the neurons' tuning curves A by U, we can derive the ensemble's functional basis. The five

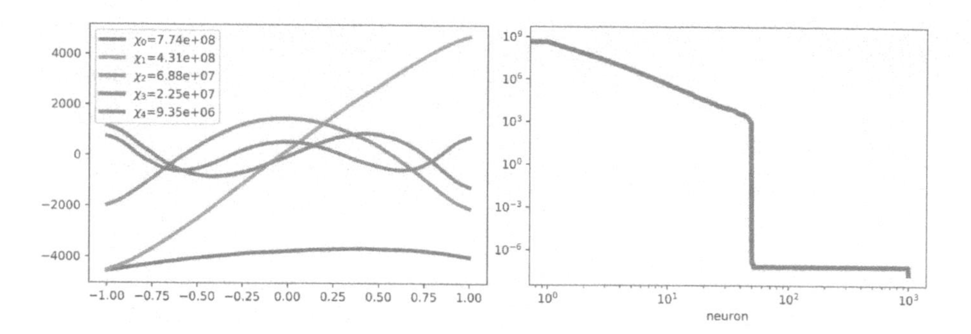

Figure 12.16 Five most important basis functions (left) and variation drop (right) for 1,000 randomly tuned neurons.

most important functional basis for a neuron ensemble with randomized tuning curves are shown in **Figure 12.16, left**. The resulted curves resemble the *Legendre polynomials* which are commonly used to approximate polynomials. Therefore, we can conclude that randomly tuned spiking neurons can represent any function which can be approximated by polynomials. By plotting S, we can get an idea regarding the rate at which the importance of each neuron's level of independence drops. For an ensemble of randomly tuned 1,000 neurons, less than 100 neurons are of high importance (**Figure 12.16, right**).

Modifying the distribution of neurons' tuning curves will change the basis functions as well as the rate of variation drop. For example, setting neurons' intercepts to be uniformly distributed across [−0.3,0.3] (as they were defined in **Figure 12.13**) will change the basis functions and the variation drop (**Figure 12.17**).

Up until now, we represented values or stimulus x in 1D. However, high-dimensional representation is often needed.

12.1.1.4 Representation of high dimensional stimulus

In a n-dimensional representation space, a stimulus is represented by n values. Particularly, in a 2D representation, each neuron is characterized with two input dimensions (x_1, x_2). Tuning curves of four 2D neurons are shown in **Figure 12.18**. In high-dimensional representation, a decoder analysis, similar to the one described above, can also be performed. Examples of basis functions for an ensemble of 500 2D neurons are shown in **Figure 12.19**.

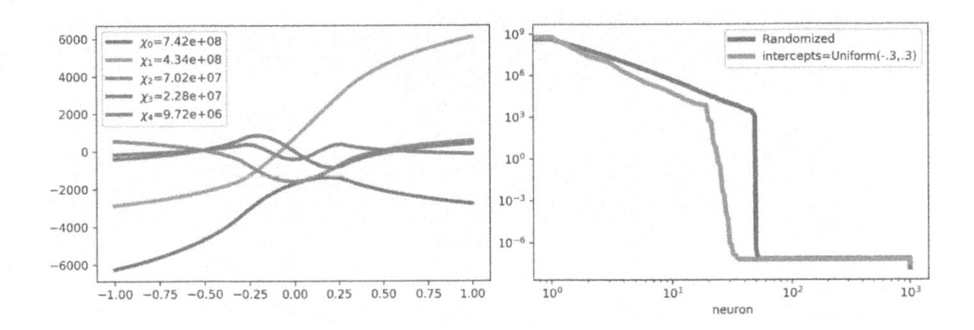

Figure 12.17 Basis functions (left) and variation drop (right) for neurons with uniformly distributed intercepts across [−0.3, 0.3], as it was derived for 1,000 randomly tuned neurons. Variation drop was compared to the one achieved in **Figure 12.16**.

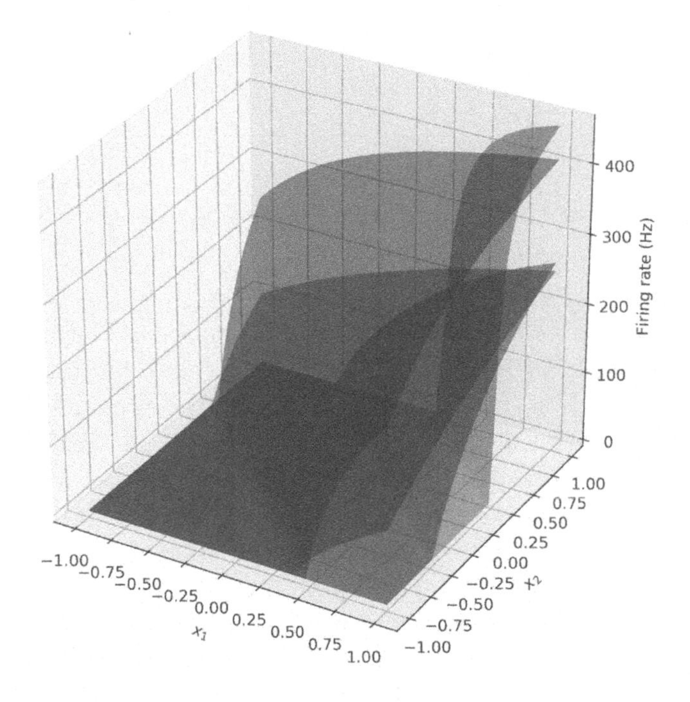

Figure 12.18 Tuning curves of four 2D neurons.

Distributing intercepts uniformly between −1 and 1 makes sense for 1D ensembles. The neuron's intercept defines the part of the representation space in which this neuron is firing. In 1D representation space, uniformly distributed intercepts uniformly span that space. A neuron with

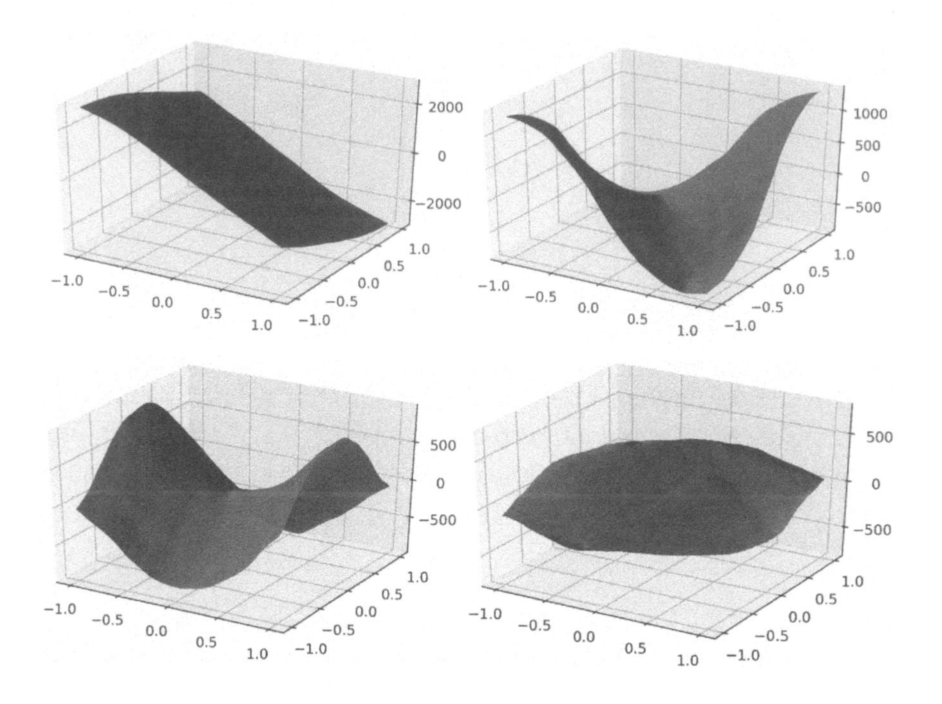

Figure 12.19 An example for of four basis functions for a 500 2D neurons ensemble.

an intercept of 0 fires spikes for 50% of the represented space ($[-1, 1]$), and a neuron with an intercept of 0.75 firing spikes for 12.5% of that space. What would be the case, however, for a 2D representation space? In a 2D space, a neuron with an intercept of 0.75 fires spikes for only $\approx$ 7.2% of the represented space. In higher dimensions, this number becomes smaller (or larger for negatively encoded neurons). In high dimensions, naively distributed intercepts lead to many neurons which are either seldom or always active. In both cases, these neurons do not contribute to the representation. An analysis plot demonstrating the activity of uniformly distributed neurons within ensembles of different dimensions is shown in **Figure 12.20**.

To better understand the high-dimensional representation space, we should analyze it more profoundly. A 2D representation space leads to a 3D tuning sphere in which each neuron's encoder points to a cap that specifies the space in which that neuron is active. The intercept is the location of the cap cutoff plane ($r - h$) (**Figure 12.21**). The ratio between the cap's and the sphere's volume is the percentage of the

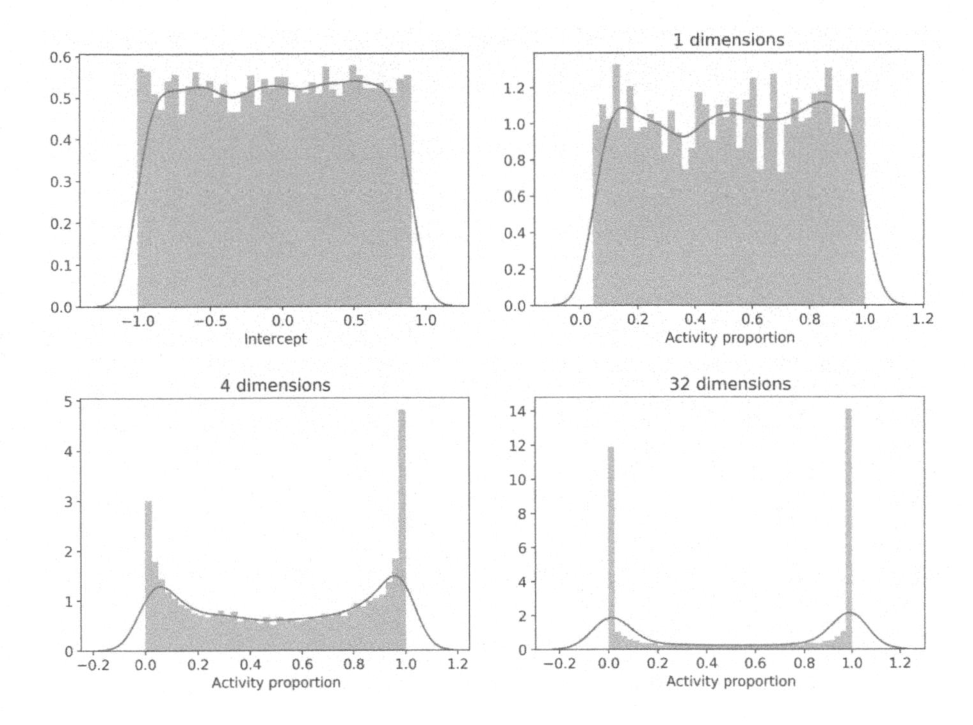

Figure 12.20 Activity analysis for uniformly distributed neurons within ensembles of 1, 4, and 32 dimensions.

representation space in which a neuron is active. A generalized sphere in a higher dimension is a *hypersphere*. The volume of a hypersphere's cap v_{cap} is defined with:

$$v_{cap} = \frac{1}{2} C_d r^d I_{\frac{2rh-h^2}{r^2}}(\frac{d+1}{2}, \frac{1}{2}) \tag{12.22}$$

where C_d is the volume of a unit hypersphere of dimension d and $I_x(a, b)$ is the *regularized incomplete beta function.*

Let x be the intercept and $r = 1$ (representation is in $[-1, 1]$). The hypersphere's cap is defined with h, where here, $h = 1 - x$. The ratio between the hypersphere's volume C_d and its cap volume v_{cap} is:

$$p = \frac{1}{2} I_{1-x^2}(\frac{d+1}{2}, \frac{1}{2}) \tag{12.23}$$

We can use the inverse of Eq. 12.23 to compute for a given value of p, the intercept which will create it. This equation is defined with:

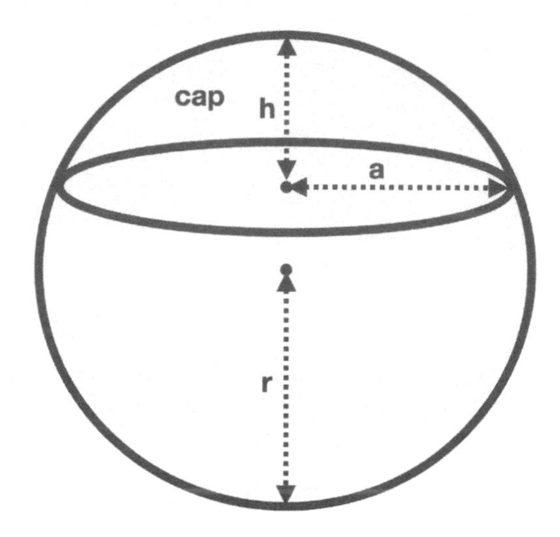

Figure 12.21 A schematic of a sphere in which a cap is designated by h and a.

$$x = \sqrt{1 - I_{2p}^{-1}(\frac{d+1}{2}, \frac{1}{2})} \tag{12.24}$$

Using Eq. 12.24, we can regenerate the intersects to better span the representation space. We can use Eq. 12.24 to define neurons as demonstrated in **Figure 12.22** (compare with **Figure 12.20**).

The analysis above was inspired by the work of *Travis Dewolf* which is succinctly summarized in [73].

12.1.2 Transformation

So far, we have discussed the neural activity in a single neuron ensemble. Transformation is the second fundamental principle udnerlying the NEF and it governs the process with which a representation of x is transformed to $f(x)$, through weighted connections between two ensembles.

12.1.2.1 Linear transformation

Considering two groups of neurons, one represents x and the other represents y. Given that each ensemble is comprised of different neurons with different tuning, how would we connect them such that $x = y$? Mathematically, encoding x with activities a, in accordance with Eq.

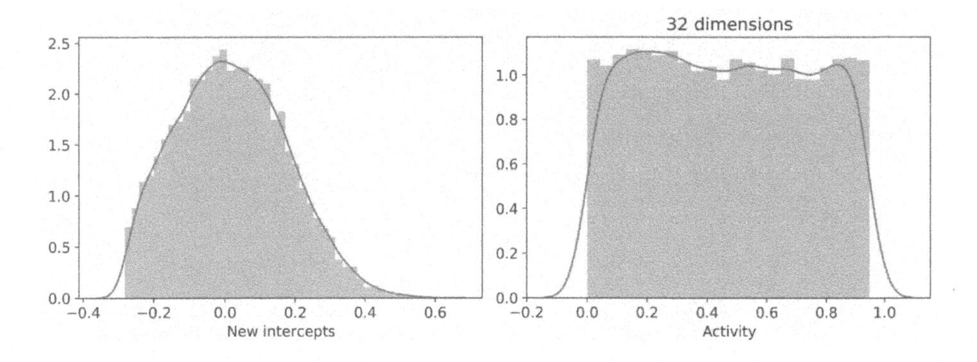

Figure 12.22 Activity analysis for redistributed neurons within an ensemble of 32 dimensions.

12.6, will take the form: $a_i = G_i[\alpha_i e_i x + J_i^{bias}]$. In accordance with Eq. 12.7, we can decode x using: $\hat{x} = \sum_i a_i d_i$. For the representation of y with a set of activities b, we will similarly define: $b_j = G_j[\alpha_j e_j y + J_j^{bias}]$ and $\hat{y} = \sum_j b_j d_j$. To achieve $x = y$, we will substitute: $y = \hat{x}$, resulting in: $b_j = G_j[\alpha_j e_j \sum_i a_i d_i + J_j^{bias}]$ which can be rearranged as: $b_j = G_j[\sum_i \alpha_j e_j a_i d_i + J_j^{bias}]$.
By defining:

$$w_{ij} = d_i \cdot \alpha_j e_j \tag{12.25}$$

we get the following definition:

$$b_j = G_i[\sum_i w_{ij} a_i + J_j^{bias}] \tag{12.26}$$

A weight matrix W which holds the multiplicative combinations of the decoders d of the first ensemble and the encoders e of the second population can be defined as:

$$W = d \otimes e \tag{12.27}$$

where the symbol $\otimes$ refers to the outer product operation defined as:

$$d \otimes e = \begin{bmatrix} d_1 e_1 & d_1 e_2 & ... & d_1 e_m \\ d_2 e_1 & d_2 e_2 & ... & d_2 e_m \\ ... & ... & ... & ... \\ d_n e_1 & d_n e_2 & ... & d_n e_m \end{bmatrix} \tag{12.28}$$

The weight matrix is calculated by multiplying the decoders from the first neuron ensemble with the encoders of the second neuron ensemble. We therefore have a nxm weight matrix, where n is the number

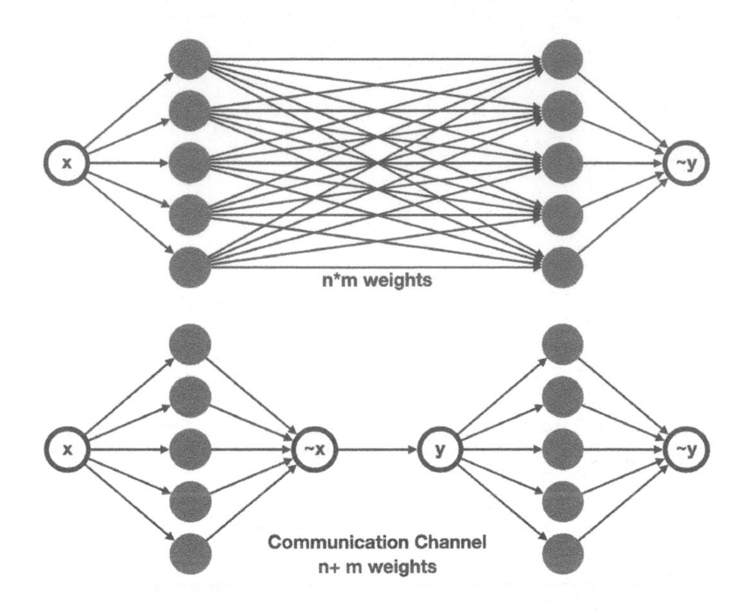

Figure 12.23 Two alternatives - numerically equivalent - transformation schemes. In the top scheme, a full nxm weight matrix is defined, while in the bottom encoding-decoding scheme, only $n + m$ multiplies are required.

of neurons representing x and m is the number of neurons representing y (**Figure 12.23**, top). As an alternative computing process, we can simply consider the straight forward encoding-decoding scheme in which only $n + m$ multipliers are required (**Figure 12.23**, bottom).

Communicating values from one ensemble to the other will not take us far. Can we transform x, to compute a function of it such that $y = f(x)$?

12.1.2.2 Linear transformations

When $f(x)$ is a linear function, the transformation is easy to calculate. It turns out that if the decoder d for x is known, a decoder for mx is md. For example, to compute $y = 2x$, we can simply multiply the "representational" decoders of x by 2 (**Figure 12.24**).

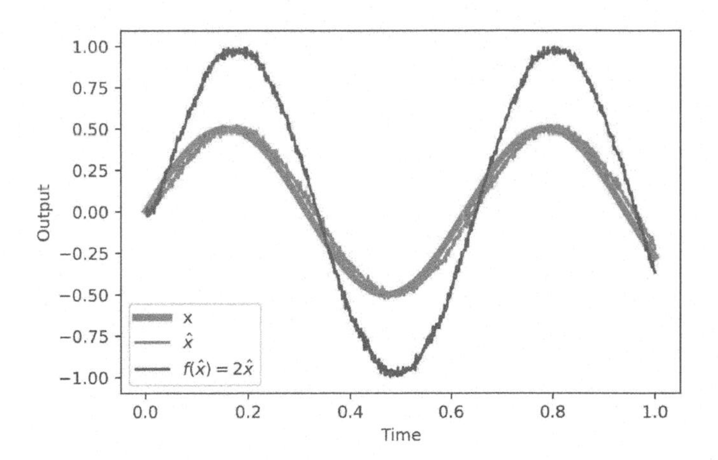

Figure 12.24 Transforming the representation of $sin(x)$ in one ensemble to $2sin(x)$ in another ensemble by scaling its decoders by 2.

12.1.2.3 Non-linear transformations

To compute a non-linear transformation of x, a different approach should be taken: using $f(x)$ when computing Υ. Particularly, Eq. 12.7 should be reformulated to:

$$\hat{f(x)} = \sum_i a_i d_i^{f(x)} \tag{12.29}$$

We changed the "representational decoders" d_i to the "transformational decoders" $d_i^{f(x)}$. In accordance with Eq.12.14, $d_i^{f(x)}$ can be calculated using:

$$d^{f(x)} = \Gamma^{-1}\Upsilon^{f(x)} \tag{12.30}$$

We substituted x with $f(x)$ when computing Υ. An example of a non-linear transformation x^2 is shown in **Figure 12.25**.

12.1.2.4 Addition

Another form of transformation is an addition: taking two values x and y which are represented by two different neuron ensembles A and B, respectively, and sum them as z by jointly connecting them to a third ensemble C. Essentially, what we are trying to do is to have the stimulus

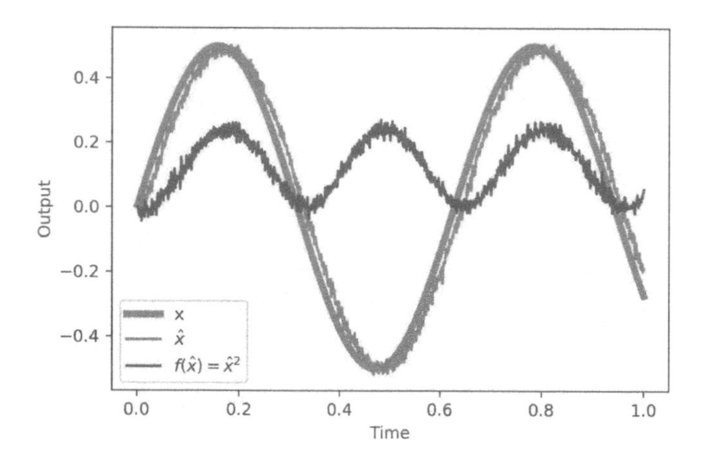

Figure 12.25 Transforming the representation of $sin(x)$ in one ensemble to $sin(x)^2$ in another ensemble by substituting x with $f(x)$ when computing Υ.

going into neuron k in the third ensemble C to be:

$$J_k = \alpha_k e_k(\hat{x} + \hat{y}) + J_k^{bias} \tag{12.31}$$

Connecting multiple inputs into a neuron automatically gives addition. Mathematically, this can be formulate as follows: following Eq. 12.7, we derive: $J_k = \alpha_k e_k(\sum_i a_i d_i + \sum_j a_j d_j) + J_k^{bias}$. This can be developed into: $J_k = \sum_i(\alpha_k e_k \cdot a_i d_i) + \sum_j(\alpha_k e_k a_j d_j) + J_k^{bias}$. Using weight notation (as was defined for Eq. 12.26) will yield:

$$J_k = \sum_i(w_{ik}a_i) + \sum_j(w_{jk}a_j) + J_k^{bias} \tag{12.32}$$

where $w_{ik} = d_i \otimes \alpha_k e_k$ and $w_{jk} = d_j \otimes \alpha_k e_k$. This computation is schematically shown in **Figure 12.26**. An example for a summation of two encoded values - each representing a shifted sin function - is shown in **Figure 12.27**. In Section 12.1.1.4, we discussed the representation of high dimensional input. We can utilize high dimensional vector representation for addition as well, as it is demonstrated in **Figure 12.28**.

12.1.2.5 Multiplication

To compute a nonlinear function of two inputs, such as: $x \cdot y$, we need to combine high dimensional representation with nonlinear transformation.

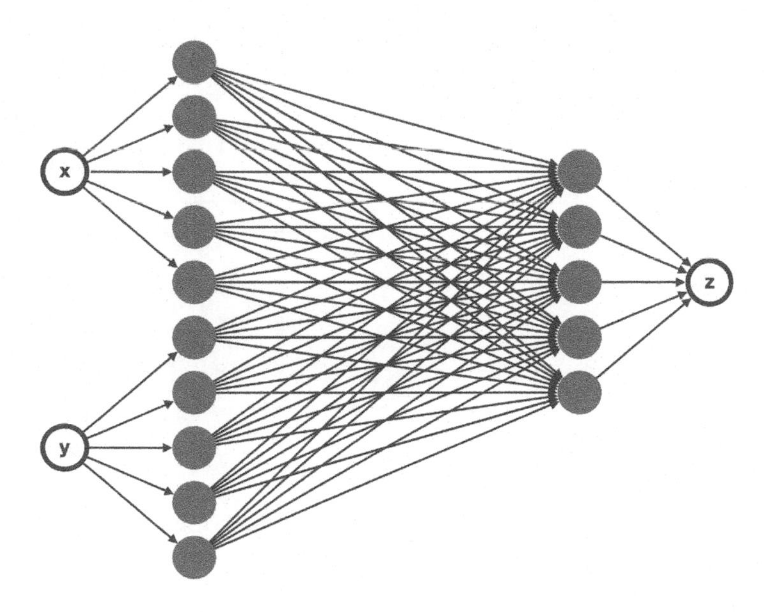

Figure 12.26 Summing two encoded values in a third ensemble with a fully connected feedforward weighted connectivity scheme.

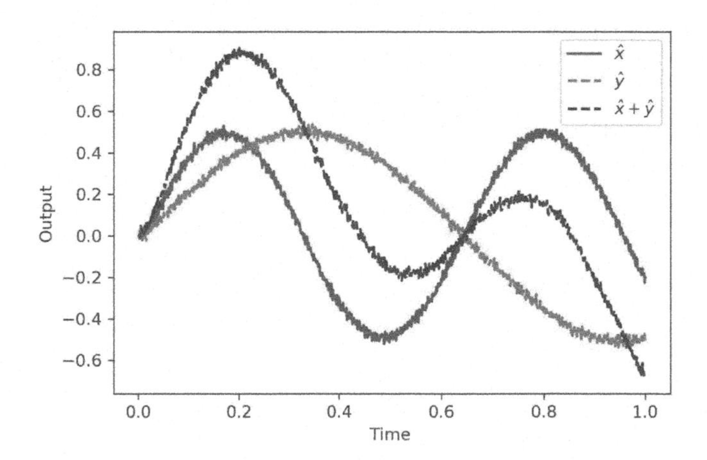

Figure 12.27 Summing two encoded values (two shifted sin functions) in a third ensemble.

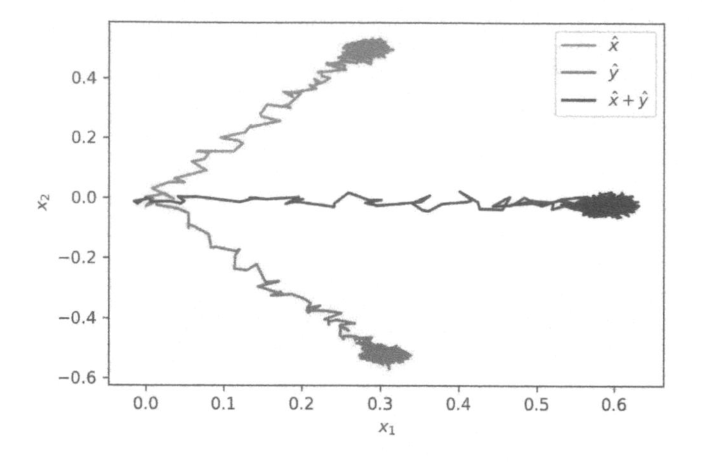

Figure 12.28 Summing two encoded vectors: $[0.3, 0.5]$ and $[0.3, -0.5]$ in a third ensemble.

In **Figure 12.25**, we demonstrated the non-linear transformation $f(x) = x^2$. If we define x as a vector in 2D space, the result of this transformation would be: $x_1^2 + 2x_1x_2 + x_2^2$. Therefore, it seems that as we combine two inputs into a 2D space, their product can be derived via optimization. An example of multiplication is shown in **Figure 12.29**. It means that 1D ensembles (or array of ensembles) cannot be used to compute nonlinear functions of more than one value.

Multiplication is a powerful tool to have and has many uses. One particularly important application is signal gating. An example of signal gating is shown in **Figure 12.30**.

In this section, we explored feedforward applications of SNNs in NEF. We used spiking activity to represent stimulus x and explored its transformation to represent $f(x)$. In the next section, we will dive into recurrent connections, allowing us to calculate differential equations.

12.1.3 Dynamics

System dynamics is a theoretical framework concerning the nonlinear behavior of complex systems in respect to time. Dynamics is the third fundamental principle underlying the NEF, and it provides the NEF with the capacity to utilize SNNs to solve differential equations. It is essentially a combination of the two first NEF principles: representation

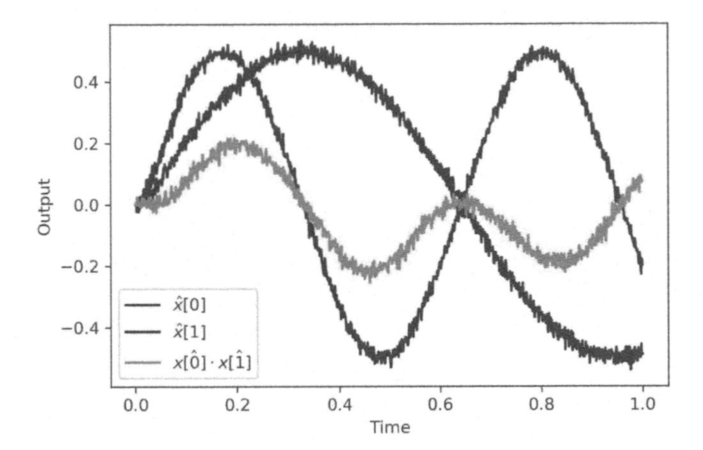

Figure 12.29 Multiplication of two encoded values in 2D in one ensemble into a third 1D ensemble.

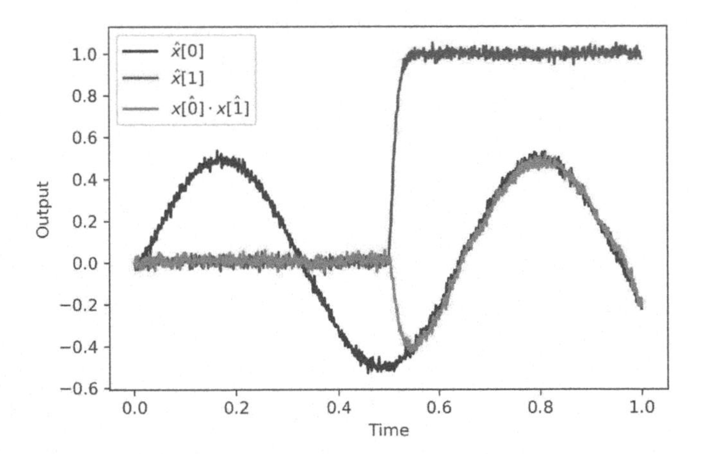

Figure 12.30 Implementation of signal gating using multiplication of an encoded value (sin) and an encoded control step signal which changes to 1 after 500 ms.

and transformation, where here, we use transformation in a recurrent connection.

12.1.3.1 *The recurrent connection*

Following Eq. 12.16, a recurrent connection (connecting a neural ensemble back to itself) is defined using:

$$x(t) = f(x(t)) * h(t) \tag{12.33}$$

To better understand this connection, which also involves convolution, we shall use the *Laplace transform*. The Laplace transform is a generalized complex version of the Fourier transform which allows the analysis of a system response to a stimulus (rather than its steady state response to a continuous signal). We can use this transform to convert the convolution operator to a multiplication using: $\mathcal{L}f(x) * g(x) = F(s) \cdot G(s)$. In our case, Eq. 12.33 becomes:

$$\begin{aligned} X(s) &= F(s) \cdot H(s) \\ H(s) &= \frac{1}{1 + s\tau} \end{aligned} \tag{12.34}$$

where s is the complex frequency parameter: $s = a + j\omega$.

Rearranging Eq. 12.34 gives $sX(s) = \frac{1}{\tau}((f(x(t)) - c(t))$. We can convert it back to time domain deriving:

$$\frac{\delta x}{\delta t} = \frac{1}{\tau}(f(x(t)) - x(t)) \tag{12.35}$$

Let's explore this result. When implementing the recurrent function $f(x) = x + 1$, we solve $\frac{\delta x}{\delta t} = \frac{1}{\tau}(x + 1 - x) = \frac{1}{\tau}$. A simulation of this model is shown in **Figure 12.31** for different values of τ. The neuron response increases until the upper representational bound is reached (defined by a *radius* parameter within Nengo).

When implementing the recurrent function $f(x) = -x$, we solve $\frac{\delta x}{\delta t} = \frac{1}{\tau}(-x - x) = \frac{-2x}{\tau}$. When the stimulus is 1, at the steady state ($\frac{\delta x}{\delta t} = 0$), the value represented by this ensemble is defined with $-2x + 1 = 0$, for which $x = 0.5$. Similarly, when the stimulus is -1, $x = -0.5$. A simulation of this model is shown in **Figure 12.32**.

When implementing the recurrent function $f(x) = x^2$, we solve $\frac{\delta x}{\delta t} = \frac{1}{\tau}(x^2 - x)$. When the stimulus is 0.2 at the steady state, the value represented by this ensemble is defined with $0 = x^2 - x + 0.2 =$

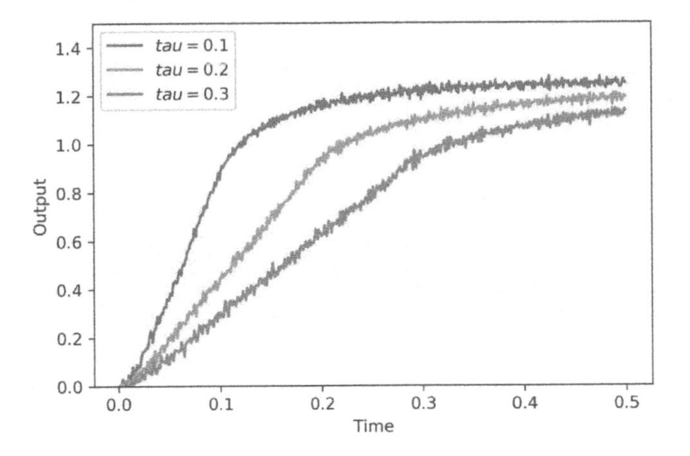

Figure 12.31 Recurrent computing of $f(x) = x + 1$ with three values of tau: 0.1, 0.2, and 0.3.

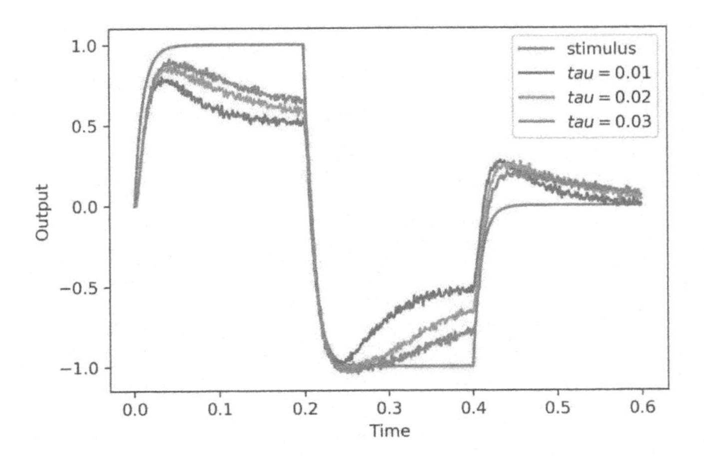

Figure 12.32 Recurrent computing of $f(x) = -x$ with three values of tau: 0.1, 0.2, and 0.3, and given a two steps ∓ 1 stimulus.

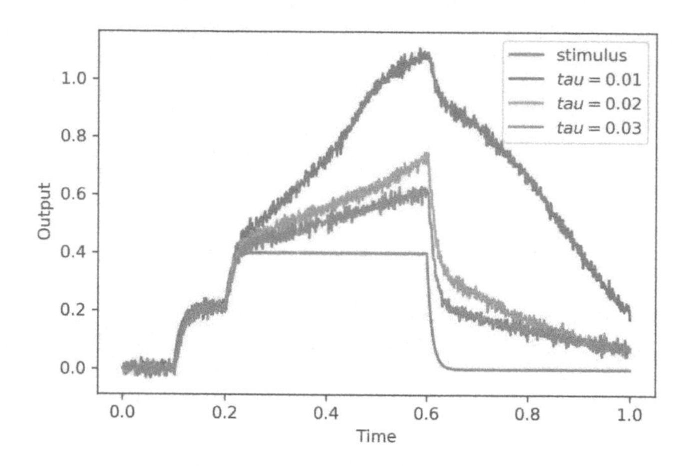

Figure 12.33 Recurrent computing of $f(x) = x^2$ with three values of tau: 0.1, 0.2, and 0.3, and given a two steps 0.2 and 0.4 stimulus.

$(x - 0.72)(x - 0.27)$, for which $x = 0.27$ or $x = 0.72$. When the stimulus is 0.4, the solution is an imaginary number. When the stimulus is 0, the solutions are $x = 0$ or $x = 1$. A simulation of this model is shown in **Figure 12.33**.

12.1.3.2 Synthesis

How can we specify a differential equation using a recurrent connection? Given a desired $\frac{\delta x}{\delta t} = f(x)$, we need a recurrent connection which, in accordance with Eq. 12.35, realizes: $\tau f(x) + x$.

If we have an input to the system $g(u)$, in accordance with Eq. 12.33, we derive: $x(t) = f(x(t)) * h(t) + g(u(t)) * h(t)$. Following the same derivation steps of Eq. 12.35, we end up with the conclusion that $g(u)$ should be scaled by τ:

$$\frac{\delta x}{\delta t} = \frac{1}{\tau}(f(x) - x + g(u)) \tag{12.36}$$

The recurrent connection should therefore realize: $\tau f(x) + x + g(u)$.

A control theory driven approach would be to reformulate these equations as:

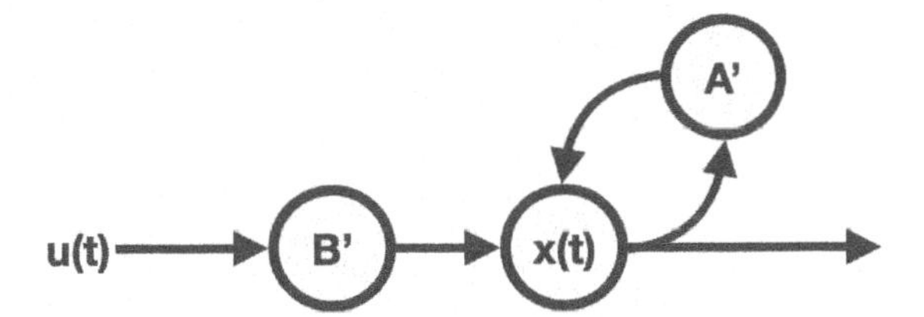

Figure 12.34 Control loop scheme in which an input u(t) is scaled by B and introduced into the system's state $x(t)$. A recurrent connection computes the transformation $f(x) = A'$. A' and B' are defined using Eq. 12.37.

$$\begin{aligned}\frac{\delta x}{\delta t} &= Ax(t) + Bu(t)\\ A' &= \tau A + I\\ B' &= \tau B\end{aligned} \tag{12.37}$$

where I is the identity matrix, A' is the function the recurrent connection is transforming, and B is the transformation applied to the input. The control schematic is shown in **Figure 12.34**.

12.1.3.3 Neuromorphic integration

An important application for NEF's third principle is the *integrator*. A simple integrator can use a velocity input signal v to derive a position x using $x = \int v$, or by solving $\frac{\delta x}{\delta t} = v$. In terms of Eq. 12.37, $A = 0$ and $B = 1$, resulting in $A' = \tau \cdot 0 + I = 1$ and $B' = \tau \cdot 1 = \tau$. A simulation of the implemented model is shown in **Figure 12.35**. The area under the velocity curve is approximately the magnitude of the position output.

A slightly more complicated integrator will incorporate a leakage with which the integrated value slowly decreases. A *leaky integrator* can be formulated with $\frac{\delta x}{\delta t} = v - \frac{1}{\tau_c x}$, where τ_c is the time constant for the leakage rate. In accordance with Eq. 12.37, we derive $A' = 1 - \tau\frac{1}{\tau_c}$ and $B' = \tau$. A simulation of the implemented model is shown in **Figure 12.36**.

Another useful version of the integrator is the *controlled leaky integrator* which has a second input d, dictating its leakage time constant. To implement the integrator, we will represent its state $x(t)$ with two

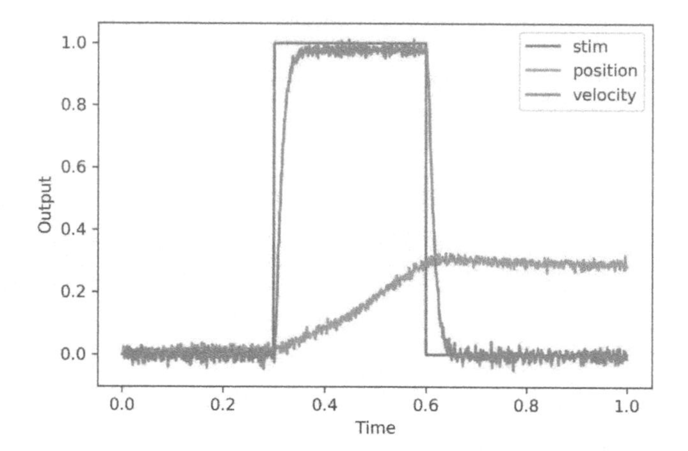

Figure 12.35 A simple integrator which takes in velocity and outputs position.

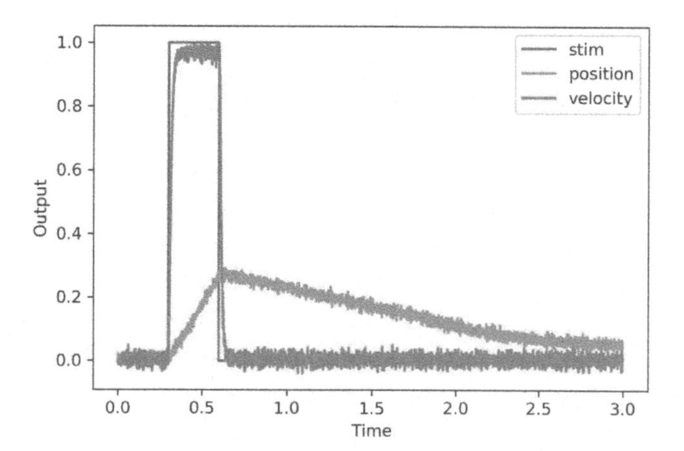

Figure 12.36 A leaky integrator which takes in velocity and outputs position with a leakage time constant of 2 sec.

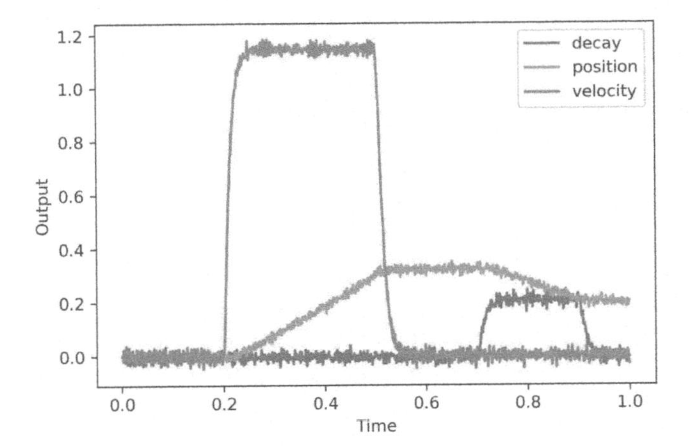

Figure 12.37 A controlled leaky integrator which takes in velocity and leakage time constant and outputs position. Here, the system initiates with no leakage ($\tau_c = 0$). For 200 msec, leakage time constant increased to 0.2, zeroing afterwards.

dimensions: one for velocity and the second for the leakage constant (see implementation details on the book website). A simulation of the implemented model is shown in **Figure 12.37**.

12.1.3.4 Neuromorphic oscillators

Oscillators are of particular interest and importance to numerous neuromorphic applications. A 2D oscillator which alternates the values represented by each of his dimensions x_0 and x_1 at rate r can be defined using:

$$\frac{\delta}{\delta t}\begin{bmatrix} x_0 \\ x_1 \end{bmatrix} = \begin{bmatrix} 1 & r \\ -r & 1 \end{bmatrix}\begin{bmatrix} x_0 \\ x_1 \end{bmatrix} \tag{12.38}$$

To achieve this behavior of having $\frac{\delta x_0}{\delta t} = rx_1$ and $\frac{\delta x_1}{\delta t} = -rx_0$, following Eq. 12.36, we need to define the recurrent connections: $x_0 = x_0 + rx_1$ and $x_1 = x_1 - rx_0$, achieving $x_0 = \frac{r}{\tau}x_1$ and $x_1 = \frac{-r}{\tau}x_0$. Implementing this model without inducing some initial value into $[x_0, x_1]$ will result in a silent oscillator: it will stand still at $[0, 0]$. However, when a stimulus is applied - even a very short one - the oscillator is driven to oscillate

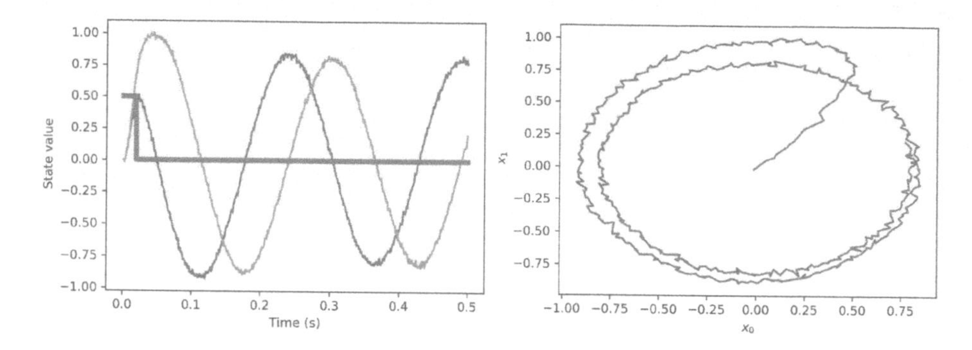

Figure 12.38 A neuromorphic oscillator, induced by a short 20 msec stimulus with $r = -0.25$. On the left panel, the values of x_0 and x_1 are shown with respect to time and, on the right panel, they are shown in a state chart.

indefinitely. A simulation of the implemented model is shown in **Figure 12.38**.

Similar to the integrator and its controlled version, we can implement a controlled oscillator. This oscillator features another dimension to incorporate adjustable frequency. A simulation of the implemented model is shown in **Figure 12.39**.

12.1.3.5 Neuromorphic attractors

An *attractor* in a dynamical system describes a certain state (or states) to which the system tends to converge. Like a ball rolling on a ground, stopping where it is most stable, a dynamical system will travel through its state-space toward a stable trajectory - the attractor. Attractors were studied in neural networks and were found to be relevant to a wide spectrum of behaviors, particularly memory [12]. The most fundamental attractor is the *point attractor* in which the system is driven to a specific state. This attractor can be simply described using: $\frac{\delta x}{\delta t} = x - p_1$ and $\frac{\delta y}{\delta t} = y - p_2$, where p_1 and p_2 constitute the designated target point in the state space. We will define the recurrent connections of a 2D ensemble in accordance with Eq. 12.35 as: $x_0 = x_0 - (x_0 - p_1)$ and $x_1 = x_1 - (x_1 - p_2)$. A simulation of the implemented model is shown in **Figure 12.40**.

The neural integrator, described above in Section 12.1.3.3 is an example of an attractor, particularly, a *line attractor*. The neural oscillator described in Section 12.1.3.4 is another example of an attractor: a *cyclic attractor*. The line attractor we have implemented above as a

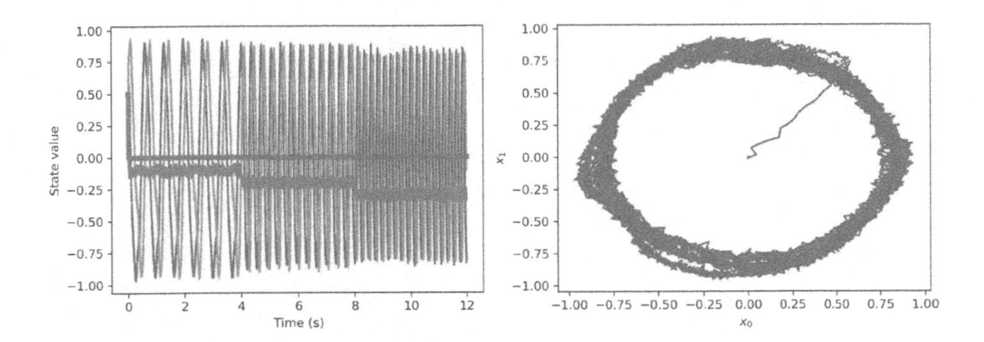

Figure 12.39 A neuromorphic oscillator induced by a short 20 msec stimulus with an adjustable rate. Rate was defined to initiate at $r = -0.1$, increased to -0.2 after 4 sec and to -0.5 afterwards. On the left panel, the values of x_0 and x_1 are shown with respect to time and, on the right panel, in a state chart.

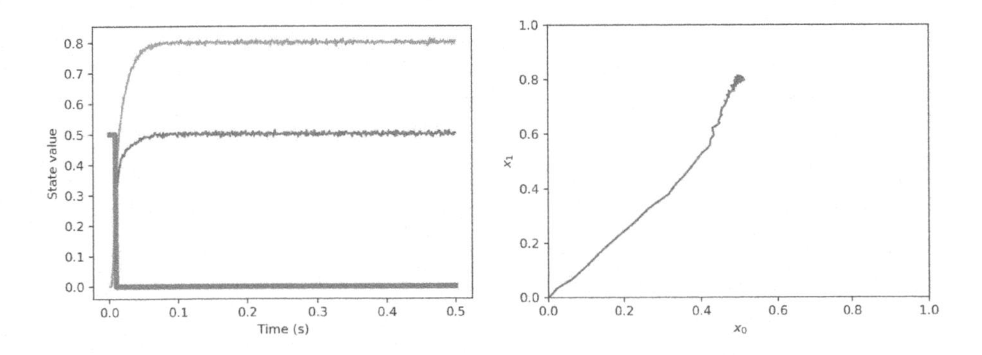

Figure 12.40 A point attractor induced by a short 10 msec stimulus and driven to point $[0.5, 0.8]$.

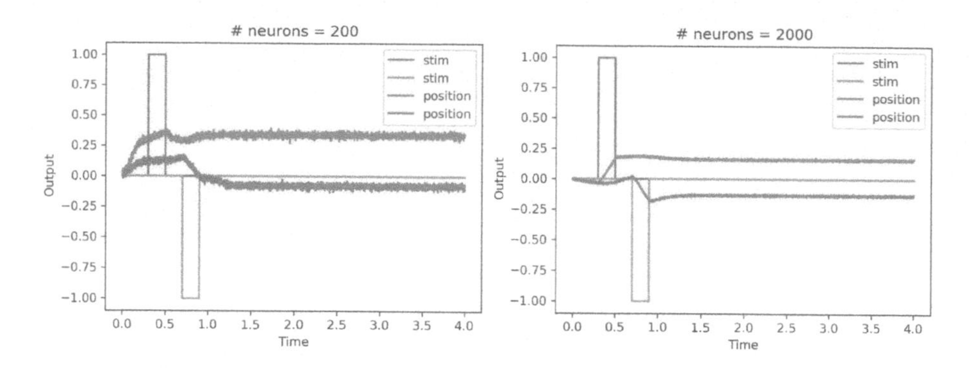

Figure 12.41 A 2D plane attractor, where each dimension was implemented as an integrator and stimulated independently.

neuromorphic integrator can be further generalized to higher dimensions. High dimensional integrators, however, require many neurons to preserve their state. Moreover, when a non-sufficient number of neurons is used, a pronounced interference effect appears: a change in one dimension affects the value represented by the second. A simulation of a 2D attractor model is shown in **Figure 12.41**.

Attractors can be used to implement *memory*. However, for high dimensional memories, we want each dimension to be independent. It will allow for accurate representation with fewer neurons and without interference effects. A simulation of the implemented model, with an array of ensembles, is shown in **Figure 12.42** (see implementation details on the book website).

An attractor can exhibit a "strange behavior." A future state of a *strange attractor* is hard to determine, as neighboring points on the attractor can be found far away later. One famous strange oscillator is the *Lorenz attractor* which has become one of the classic icons of modern nonlinear dynamics. With its intriguing double-lobed shape and chaotic dynamics, the Lorenz attractor has symbolized order within chaos [265]. The attractor can be defined in 3D as:

$$\frac{\delta x}{dt} = \begin{bmatrix} -\sigma x_0 + \sigma x_1 \\ -x_0 \cdot x_2 - x_1 \\ x_0 \cdot x_1 - \beta(x_2 + \rho) - \rho \end{bmatrix} \tag{12.39}$$

where σ, β, and ρ are the attractor parameters. In accordance with Eq. 12.36, we will define the following recurrent connections for the 3D

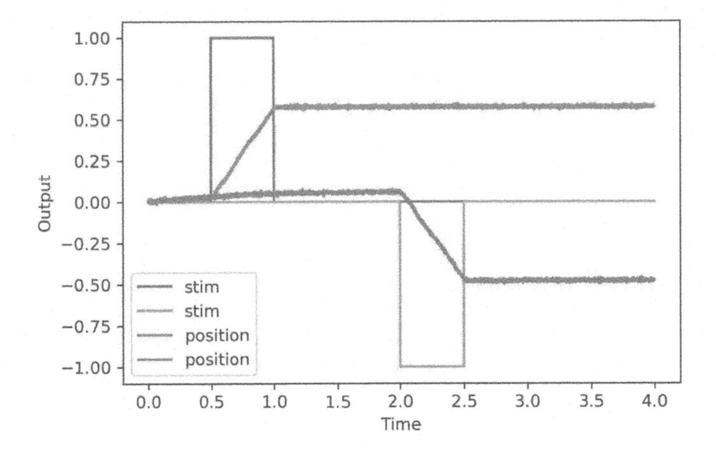

Figure 12.42 A plane attractor, implemented with an array of two neuron ensembles where each ensemble was implemented as an integrator and stimulated independently.

ensemble: $x_0 = x_0 + \tau\frac{\delta x_0}{dt}$, $x_1 = x_1 + \tau\frac{\delta x_1}{dt}$ and $x_2 = x_2 + \tau\frac{\delta x_2}{dt}$. Results for the implemented Lorenz attractor are shown in **Figure 12.43**.

12.2 CASE STUDY: MOTION DETECTION IN A SPIKING CAMERA USING OSCILLATION INTERFERENCE

This section will showcase one application for NEF-based neuromorphic oscillators, particularly, the derivation of motion detection from a visual field, reported by a spiking camera. A full description of this section is given in [285]. We will use this opportunity to briefly introduce spiking cameras (or Dynamic Vision Sensors (DVS)), one of the most important neuromorphic hardware developments. This case study is comprised of four parts:

1. **Spiking camera**. In this case study, we represent the visual field with a spiking camera which reports changes in luminance as spikes. Therefore, the input data to our model are spikes, representing motion in the visual field.

2. **Gabor functions**. Gabor functions are filter kernels inspired by the primary visual cortex (V1) which can extract the locations and directions in space where the utmost changes in intensity appear

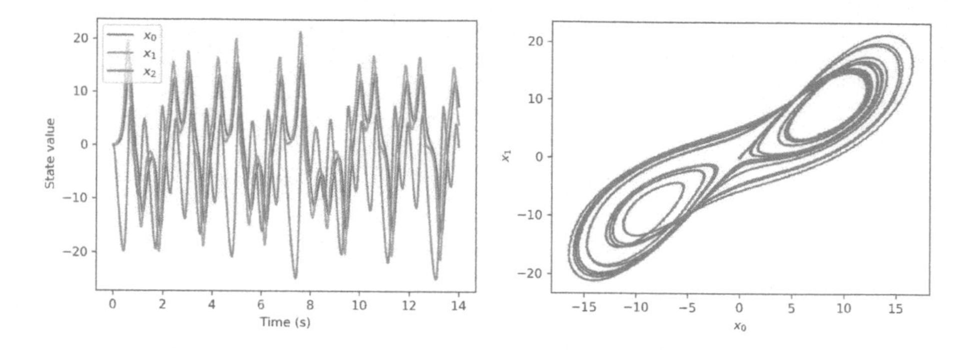

Figure 12.43 Lorenz attractor defined with $\sigma = 10$, $\beta = \frac{8}{3}$, $\rho = 28$ and $\tau = 0.1$. On the left panel, the values of x_0, x_1, and x_2 are shown with respect to time and, on the right panel, the values of x_0 and x_1 are shown in a state chart.

[149]. We will use these functions to design direction-selective oscillators.

3. **Damped oscillators**. An oscillator which is driven back to its origin in accordance with a damping factor. Here, damped behavior will allow real-time monitoring of a direction change.

4. **Motion detection**. By utilizing Gabor functions and damped oscillators, a motion detector can be designed.

12.2.1 Spiking camera

Spiking cameras were briefly introduced in Section 2.3.3. Spiking data, generated by a DVS from an animation of vertically drifting bars, is shown in **Figure 12.44**. The visual scene is converted to spiking events and communicated with the AER protocol, introduced earlier in Section 9.1.

12.2.2 Gabor functions

In this case study, we will use Gabor functions to define direction selective damped oscillators. Gabor functions are mathematically interpreted as an exponentially decaying 2-dimensional sinusoidal wave $G(x, y)$ and are specified with:

$$G(x,y) = exp(\frac{x_d^2 + \xi^2 y_d^2}{2\sigma^2}) cos(2\pi\kappa x_d + \phi) \tag{12.40}$$

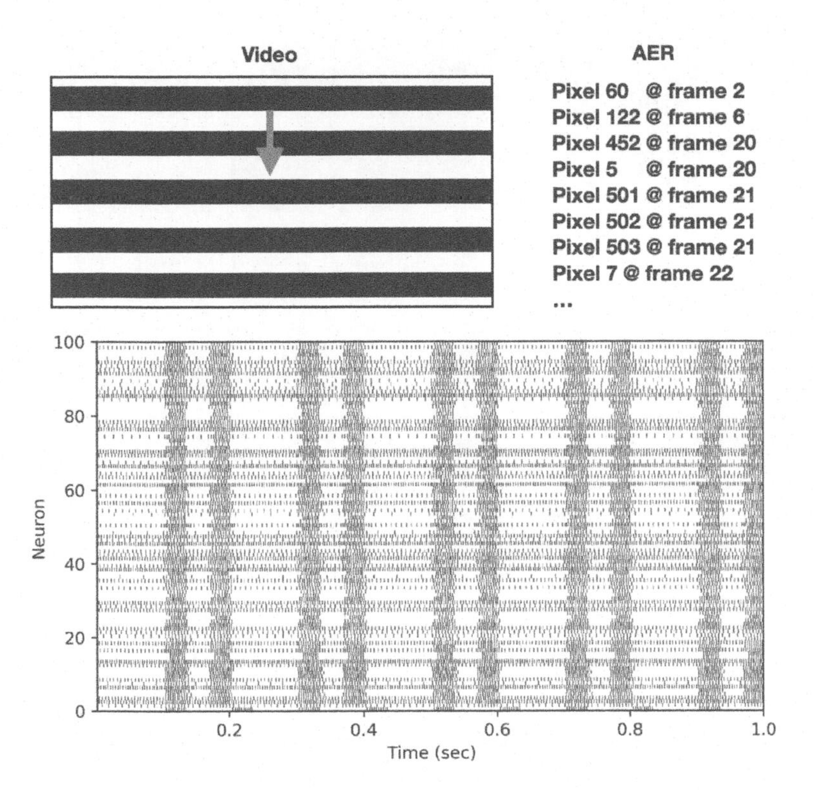

Figure 12.44 Spiking camera, where a visual scene of vertically drifting bars is represented with events, reporting changes in luminance with the AER protocol.

where σ is the Gaussian width, and ξ, ϕ, κ are the function's ellipticity, phase, and wave-number, respectively. Here, we will define the Gabor function with: $x_d = xcos(\theta) + y\sin(\theta)$ and $y_d = ycos(\theta) - xsin(\theta)$, where θ is the Gabor orientation. This formulation is similar to the elliptic 2D Gabor function proposed in [149], excluding the decay parameters (originally defined separately for each spatial dimension). This simpler Gabor form is sufficient as the functions decay similarly in both dimensions.

We would like to use Gabor functions to perceive a moving edge across the visual field. Gabor's orientation and phase define the orientation and movement direction of that edge. We will define eight Gabor functions with eight different orientations to allow a perception of motion in horizontal, vertical, and diagonal directions (**Figure 12.45**). The

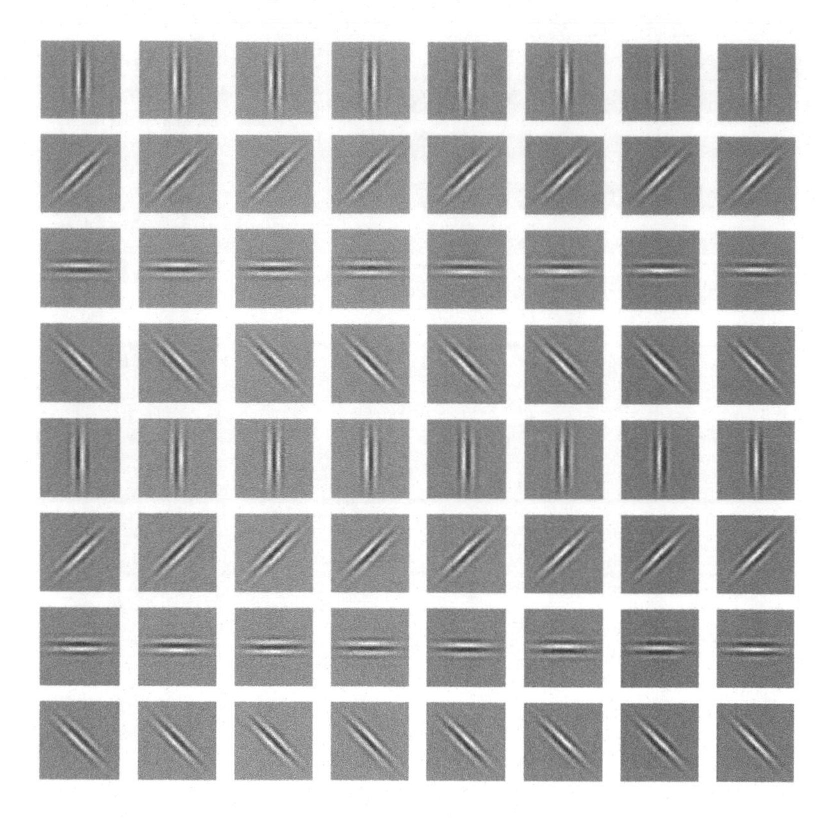

Figure 12.45 A grid of Gabor filters in which in each row and column, Gabor's orientation and phase are shifted by 45 degrees, respectively.

selectivity for edge size can be optimized for various shapes and set by the Gabor wave-number (the inverse of the sinusoidal wavelength).

12.2.3 Damped oscillators

A 2D oscillator which alternates two values: x_0 and x_1, at a rate r, was defined earlier in Eq. 12.38. Here, however, we would like to design an oscillator which will be driven back to its origin following a damping rate λ, as long as it is not induced by some stimulus. A damped oscillator can be defined using:

$$\frac{\delta}{\delta t}\begin{bmatrix} x_0 \\ x_1 \end{bmatrix} = \begin{bmatrix} 1 & r \\ -r & 1 \end{bmatrix} - \begin{bmatrix} \lambda & 0 \\ 0 & \lambda \end{bmatrix}\begin{bmatrix} x_0 \\ x_1 \end{bmatrix} \tag{12.41}$$

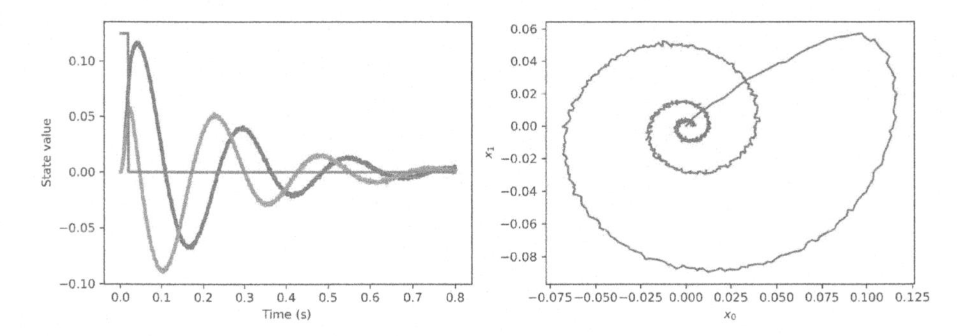

Figure 12.46 A damped oscillator defined with $\lambda = -4$, $r = 8\pi$, and $\tau = 0.1$. On the left panel, the 20 msec stimulus is shown in red and the values of x_0 and x_1 are shown with respect to time. On the right panel, x_0 and x_1 are shown in a state chart.

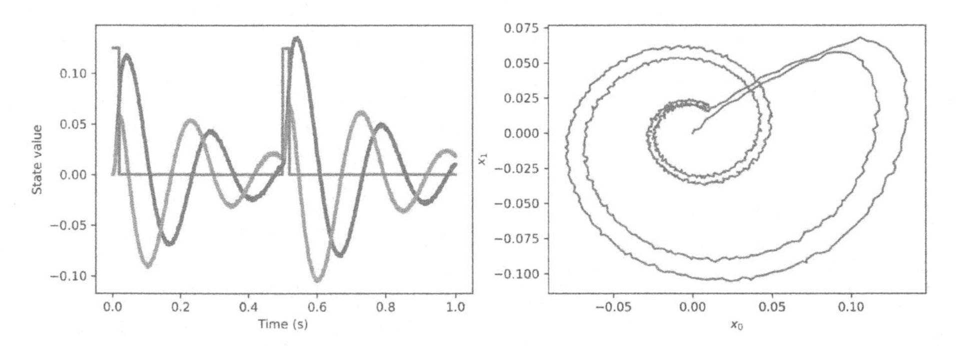

Figure 12.47 A damped oscillator defined similarly as in **Figure 12.46**, where two 20 msec stimuli were induced, in $t = 0$ and $t = 0.5$.

An example of a damped oscillator shortly induced at $t = 0$ is shown in **Figure 12.46**. In accordance with Eq. 12.36, we will define a recurrent connection with: $x_1 = \tau\lambda x_1 + x_1 + \tau r x_2$ and $x_2 = -\tau r x_1 + x_2 + \tau\lambda x_2$.

This oscillator's powerful aspect, which is of particular interest here, is that oscillation can be explicitly re-induced, as is demonstrated in **Figure 12.47**. In this example, the oscillator is induced to oscillate for one round; it damps back to zero and then induced again by a similar stimulus. In the next section, we will see how this property can be used to detect motion; we will re-induce an oscillator into action when we detect motion in a particular direction, thus defining this oscillator as direction selective.

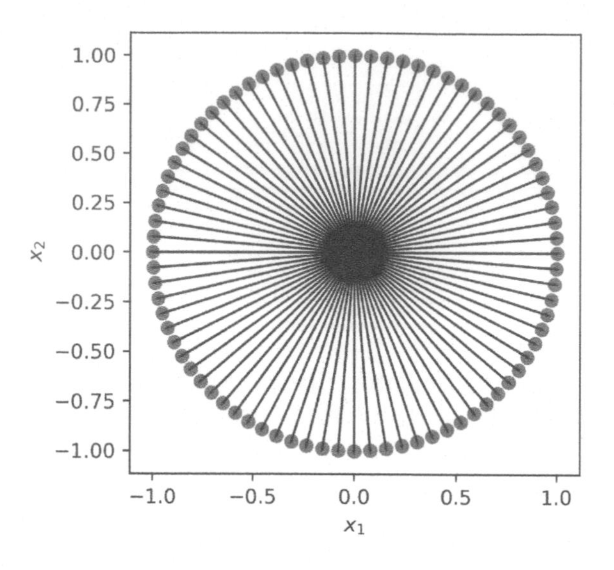

Figure 12.48 80 phase-shifted encoders in 2D were defined for each oscillator, matching the 80 phase-shifted Gabor functions.

12.2.4 Motion detection

Motion in the visual field can be represented using Gabor functions with different parameters. Above, we defined eight Gabor functions with equally-spaced orientations along the unit circle. For motion detection, we will use each Gabor function as a spatial filter for selectively driving a damped oscillator when a visual pattern matches the filter's orientation. While a stationary edge will activate the relevant oscillator, the oscillator will be driven back to origin as long as it is not induced at the appropriate time. To obtain motion detection, each Gabor function was phase-shifted at multiple phases, such that a moving pattern will match a series of Gabor functions and sequentially activate and reactivate the oscillator. To implement it, for each oscillator we have assigned 80 phase-shifted Gabor functions, equally-spaced along the unit circle. Each oscillator was therefore defined with 80 encoders (**Figure 12.48**), each aligned with a Gabor function.

A direction selective oscillator, tuned via its Gabor functions to detect downwards vertical motion, will be induced into an oscillatory state only when such motion is detected. The oscillatory pattern is summed to represent the oscillator activation state: was motion detected or not? By summing and multiplying the oscillators' state with their preferred

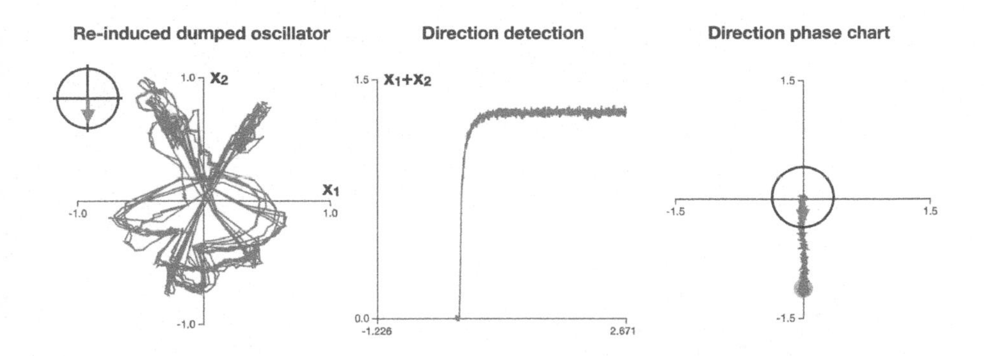

Figure 12.49 A direction selective oscillator, tuned via its Gabor functions to be induced into an oscillatory state, when vertical motion is detected. The oscillatory pattern is summed to provide a point attractor. Only the active oscillator is shown.

direction (a 2D vector), a global direction detection ensemble can be defined. This direction detection ensemble will act as a point attractor, pointing at the detected motion direction. An example for such a system is demonstrated in **Figure 12.49**, where a network of eight oscillators was induced by a downward vertical motion (shown in **Figure 12.44**), and only one of them responded, per its tuning (only the active oscillator is shown; see the book website for implementation details).

In [285], networks of oscillators were defined to span an entire 2D visual field. It was implemented by translating the Gabor functions to different locations and defining hierarchical motion detection cells in which local direction detectors are averaged, reducing noise and highlighting the differences between local and global motion.

12.3 GLOSSARY

NEF-based dynamics: The manifestation of a dynamical system from recurrently connected neuron ensembles.

NEF-based representation: Distributed representation of mathematical constructs using nonlinear - spike-based - encoding and linear decoding.

NEF-based transformation: The transformation of one representation to another though weighted synaptic connections between two neuron ensembles.

Nengo: A Python-based software library which provides an API to NEF functionality, thus allowing the simulation of large-scale neuronal models.

Neural engineering framework: A framework for the representation and transformation of high dimensional mathematical constructs with SNNs.

12.4 FURTHER READING

- **Section 12.1**
  - The main reference, outlining the fundamental principles of NEF is the book "Neural engineering: Computation, representation, and dynamics in neurobiological systems," [84]. NEF was extended to support the Semantic Pointer Architecture (SPA) which is described in the book "How to build a brain: A neural architecture for biological cognition" [83].
  - A technical overview of NEF is given in [266] and it is succinctly described in [267].
  - Implementation details of NEF-based models using Nengo is available at nengo.ai/nengo.

CHAPTER 13

Learning spiking neural networks

Abstract

Learning SNNs are of particular interest across all three perspectives on neuromorphic engineering. In Section 7.4, learning was discussed from the scientist's perspective. Here, we will revisit learning, discussing it from the perspective of the algorithm designer. We will explore unsupervised and supervised learning rules, transfer learning from artificial to spiking networks, and finally, examine neuromorphic learning as it was implemented in Nengo with an example discussing the classification of handwritten digits using NEF-based convolutional SNN.

13.1 INTRODUCTION TO LEARNING SNNS

From the perspective of the scientist, learning is key to cognition (Sections 7.4 and 3.1). Biological learning, however, is also one of the greatest inspirations for algorithm designers.

Section 7.4 concluded with a short description of STDP, described as a general unsupervised learning mechanism in which synapses undergo LTP or LTD. These learning rules are based on Hebbian learning, and they feature *local* learning, thus allowing biologically plausible learning which is not orchestrated by central processing units. STDP is one of the most utilized biologically plausible learning strategies in neuromorphic systems. It is mainly used for real-time learning where synaptic weights are continuously updated following pre- and post-synaptic spikes' relative timing. Relative timing constitutes a causal relationship between events. Learning therefore takes place in particular time windows in which causal relationships between neurons take place [226].

These learning rules provide a neuromorphic means for unsupervised learning in which the learning objectives are not known. However, they were shown to be useful learning techniques. For example, within a neuromorphic CNN, they were used for unsupervised extraction of visual features as well as for image classification [156]. CNN is a DNN in which learning takes place in convolutional layers. A convolutional layer comprises several neuronal maps; each detects the same visual feature (e.g., edge) at different places across the input. This *weight sharing* strategy of convolutional layers makes CNNs efficient in regard to the number of trainable weights. For example, when a spiking CNN was introduced to natural images and driven by STDP learning, neurons' activity was tuned to represent prototypical visual patterns (e.g., edges). Learning was initiated with randomized, normally distributed weights. Therefore, ahead of training, the neurons are not yet selective to specific patterns. As learning progresses, they gradually responded stronger to apparent visual patterns. When a new image was introduced to the network, neurons competed with each other, such that those who fired earlier triggered STDP and learned the input pattern (global intra-map competition). A visual feature is, therefore, integrated data extracted layer by layer.

Machine learning predominantly relies on supervised learning in which learning objectives are known and examples of required behavior are given (see Section 3.1). One of the classic supervised learning rules is the *Widrow-Hoff learning rule* [303] which modifies synaptic weights, such that an error signal is minimized. It is sometimes referred to as the *Delta rule* or the Least Mean Square (LMS) rule [209]. The most widely utilized way to employ the Widrow-Hoff learning rule is through gradient-descent, a predominant learning mechanism for ANNs. One version of it which was designed for SNNs is SpikeProp [44] in which a target spike train is fixed at the output layer. SpikeProb and its numerous variants have been demonstrated as comparable to backpropagation-based ANNs [248]. Supervised learning can be implemented with the NEF using the PES [294] learning rule, which will be discussed later in this chapter.

RNNs, ANNs with feedback connections; are widely used for sequential pattern recognition in text, audio, and video. Feedback allows updating the network's state according to both input data and its previous state. Naive RNNs can handle short-term data dependencies quite well; however, they often fail to relate temporally distant data. To allow learning with long-time dependencies, Long Short-Term Memory (LSTM) network architecture was developed. Most state of the art temporal sequential patterns recognition (e.g. language modeling [271] and sentiment analysis [299]) were achieved with LSTMs. In LSTMs, the network's previous state is conveyed forward while being affected by incoming information through regulating gates (alleviate vanishing and exploding gradients which are commonly associated with RNNs). Numerous variants to these gates were proposed, including a forget gate and the introduction of peephole connections. Modern architecture can identify temporal patterns across thousands of time-steps. LSTMs are the focal points for many books and they are not the main focus here. They are succinctly reviewed in [310]. Not surprisingly, BNNs have an incredible capacity for temporal pattern recognition as biological signals corresponding to a continuously changing environment. LMU is a recurrent architecture designed for SNNs, allowing continuous network optimization, even for infinitely small time steps [296]. LMUs were shown to handle temporal patterns across hundreds of thousands of time steps while achieving superior performance over LSTMs. They might become the most important application of SNNs.

All of these grand achievements have the potential to improve the utilization of SNNs in real-world applications. In the following few sections, PES learning rule, a supervised biologically plausible perceptron-based learning for SNN, will be discussed.

13.2 LEARNING SPIKING NEURAL NETWORKS WITH NEF

In many of the examples in previous chapters, we saw how future performance could be affected by past activity. For example, in Section 12.1.3.3, we used an integrator to derive position from velocity. Can this be qualified as learning? Here, we will refer to this type of dependent behavior as adaptation. However, learning is more traditionally defined by the modulation of weighted connections between neuron ensembles during optimization.

How is learning different from what was described earlier in Section 12.1.1.2, where we discussed decoding as a method to derive optimal connections' weights for the representation, or transformation, of encoded values? What is, therefore, the benefit of learning? Learning can help when the function we want to represent is unknown or when a desired behavior might change over time.

13.2.1 The prescribed error sensitivity rule

Traditionally, as was described in Section 13.1, learning is the means by which neural networks are constructed. Usually, learning is guided incrementally through some optimization method (most prominently, gradient descent). The classic perceptron learning [246] (briefly introduced in Section 3.1) is governed by the delta rule which formalizes the change in weight in accordance with some error. Assuming neurons are driven through a linear activation function, for a neuron j, the delta rule for its ith weight w_{ji} is given by:

$$\Delta w_{ji} = \alpha(t_j - y_j)x_i \tag{13.1}$$

where t_j is the target output, y_j is the given output, and x_i is the ith input.

A similar biologically plausible delta rule can be derived and implemented with NEF, as was demonstrated in [186]. In Eq. 12.9, we derived an error E for the derivation of optimal decoders. We will reformulate this expression here as:

$$\frac{\delta E}{\delta d_i} = -\int_{-1}^{1} (x - \sum_{j}^{n} a_j d_j) a_i \, dx \tag{13.2}$$

In NEF notation, Eq. 13.2 can be re-written to closely resemble the delta rule:

$$\Delta d_i = -\kappa(x - \hat{x})a_i \tag{13.3}$$

Decoders however are not realistic. In contrast to weights, they do not exist. We should therefore refer to Section 12.1.2, where we described transformational decoders in terms of weights using Eq. 12.25:

$$w_{ij} = \alpha_j d_i \cdot e_j \tag{13.4}$$

From Eq. 13.3 and 13.4, we can define:

$$\Delta w_{ij} = \alpha_j \kappa(x - \hat{x})a_i \cdot e_j \tag{13.5}$$

By defining $E = x - \hat{x}$, we conclude:

$$\Delta w_{ij} = \kappa a_i(\alpha_j E \cdot e_j) \tag{13.6}$$

The term $\alpha_j E \cdot e_j$ is the stimulus that this neuron would get if it had E as an input. However, with PES, E modulates connections weights and

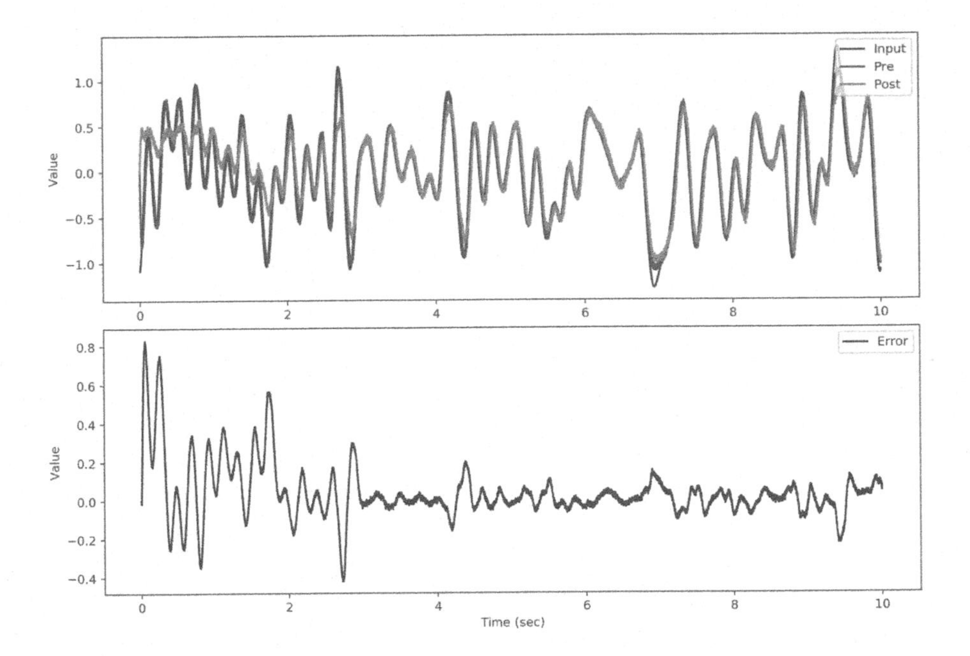

Figure 13.1 PES-optimized communication channel between one ensemble (*pre*) to another (*post*), where the input is white noise.

not driving neurons. Eq. 13.6 is the PES rule. It is biologically plausible as it uses locally available error signal.

To demonstrate PES-based learning, we will consider a randomized communication channel between one ensemble (*pre*) to another (*post*), where the input is white noise.

A third ensemble (*error*) will connect the pre and post ensembles to it, where the latter is transformed by −1. Thus, ensemble *error* is encoding the value $pre - post$. By connecting the error ensemble to the communication line which connects the *pre* to the *post* ensemble and defining it as a driver for PES learning, we initiate learning. Following few iterations, the learning rule modulates the weights of the communication line, such that the values represented by the *post* ensemble closely follow the values represented by the *pre* ensemble (**Figure 13.1**). Implementation details are given on the book website.

The PES rule was shown to be able to learn any linear and non-linear functions. A more profound overview of PES-based learning is available in [294].

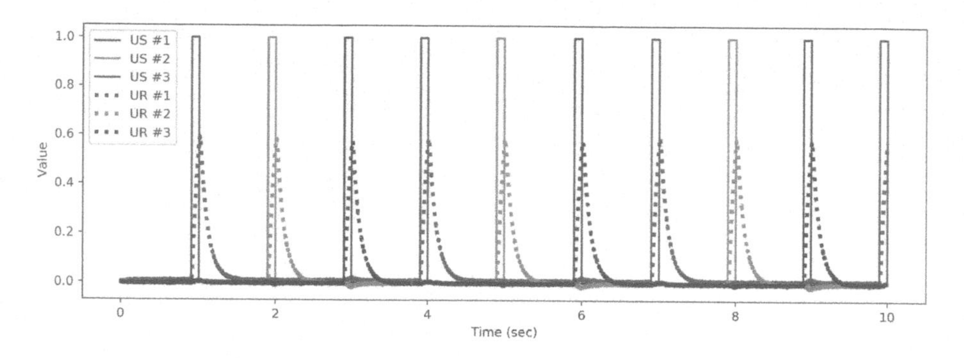

Figure 13.2 Three coupled unconditional stimuli and unconditional responses.

13.2.2 PES learning for classical conditioning

Classical or Pavlovian conditioning uses an Unconditioned Stimulus (US) (e.g., food for a dog) that elicits an Unconditioned Response (UR) (e.g., salivating) to cause a Conditioned Response (CR) (e.g., salivating after learning) to be elicited by a Conditioned Stimulus (CS) (e.g., ringing a bell). Pavlovian conditioning was first studied by *Ivan Pavlov* back in 1897 and became one of the most studied phenomena in Psychology [243]. In wave-forms, the coupling between three USs and URs is visualized in **Figure 13.2**. The UR is elicited from the US and this was implemented in NEF by connecting two 3D ensembles, where each US-UR coupling is represented in a different dimension. Implementation details are given on the book website.

One of the most well known models for Pavlovian conditioning is the *Rescorla-Wagner model* [259], which states:

$$\Delta V_x = \alpha(\lambda - \sum_x V) \tag{13.7}$$

where V_x is the value of the CS x, α is the learning rate and salience parameter, and λ is V's max value (usually 1). The term $(\lambda - \sum_x V)$ is the reward prediction error. For simplicity, we will create a model with only one element in $\sum V$, assuming there is little association between other stimuli and the US.

We will define three CSs and couple them to URs by activating them simultaneously. To make it a little more interesting, for each UR, we will generate three CSs with different magnitudes. Thus, generating a CR following the largest CS to which it was coupled. This coupling

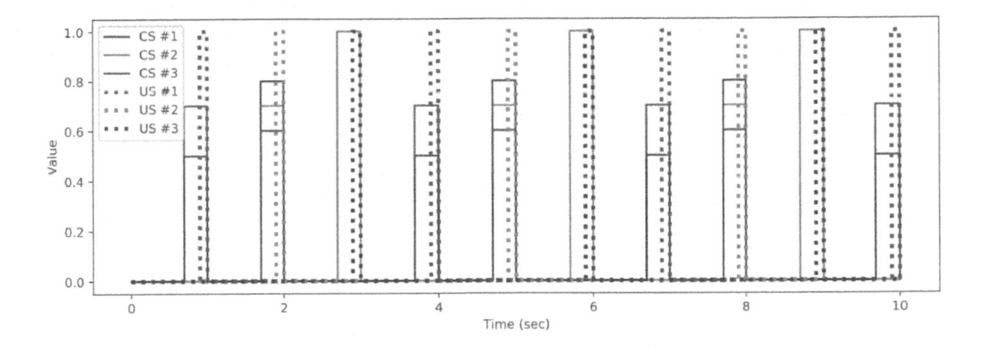

Figure 13.3 Three coupled conditional stimuli and unconditional responses. The blue CS is coupled with the blue UR, the black CS is coupled with the red UR, and the red CS is coupled with the black UR.

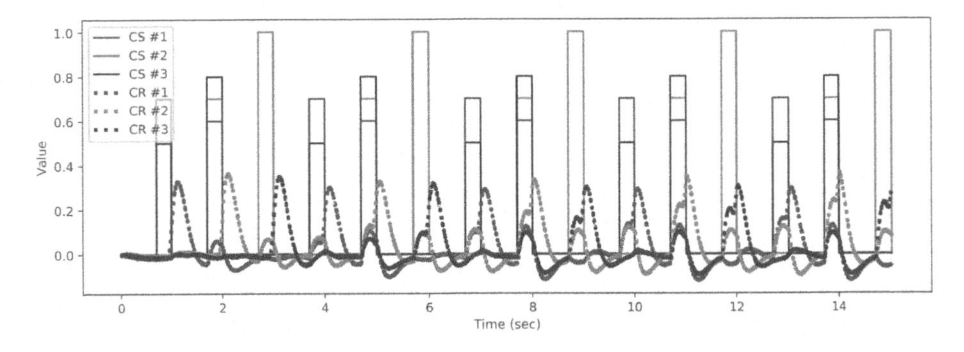

Figure 13.4 Three conditional responses generated by the corresponding conditional stimuli.

scheme is visualized in **Figure 13.3**. By activating learning, we can examine the generated CR by the corresponding CS. Results are shown in **Figure 13.4**. It seems that a CR is generated proportionally to the CS's magnitude to which it was coupled.

To further investigate the learning methodology in Pavlovian conditioning, we introduced a node through which learning is inhibited in the first and last few seconds. Results are shown in **Figure 13.5**. In the beginning, when learning is off, CSs do not elicit CRs. When learning is activated, CRs are generated as expected. When learning is turned off again, CRs are generated but in a much less regulated pattern.

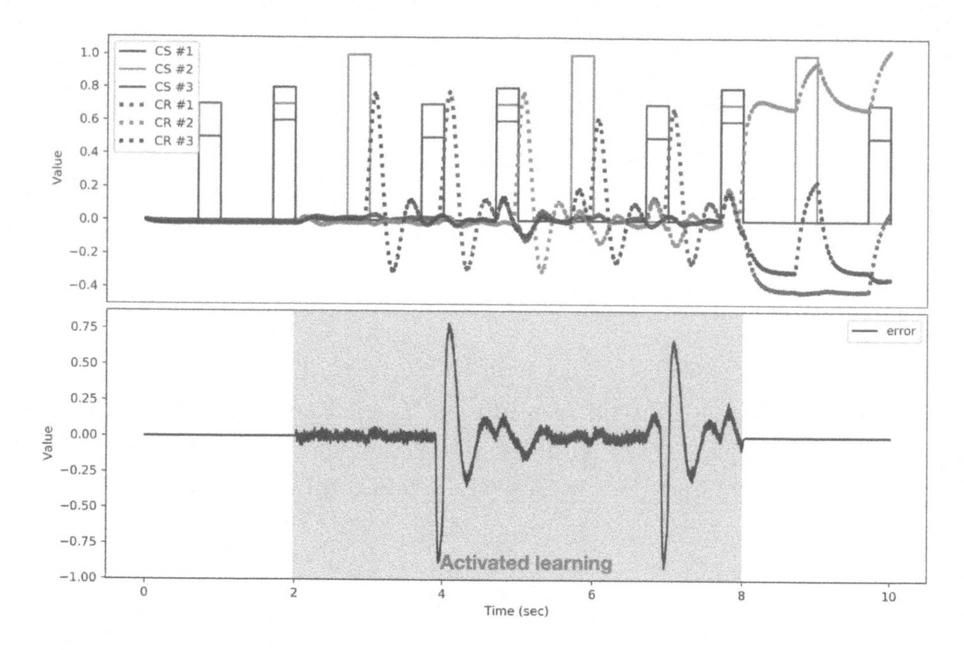

Figure 13.5 Three conditional responses generated by the corresponding conditional stimuli.

13.3 FROM DNN TO DEEP SNN

As was demonstrated by *Eric Hunsberger* and colleagues, converting a DNN to a deep SNN is a straight forward task with which a SNN can exhibit state-of-the-art results [134].

With Eq. 5.23, given a constant input current, a neuron's steady-state firing rate can be calculated. As was shown in Section 12.1.1.2, with NEF, temporal filtering is applied to spiking neurons. Assuming that synaptic filtering smoothes a spike train to give a good approximation of the firing rate, LIF neurons can be used for optimizing a network with traditional gradient descent. However, as we know from examining a neuron's tuning curves (Section 12.1.1.1), the function derivative approaches infinity at the intercept. Nondifferentiability is problematic in gradient descent which relies on the propagation of errors through differentiable mathematical constructs. To provide differentiable LIF neurons, we can modulate our existing model to feature smoother, softer transitions.

Equation 5.23 can be slightly rearranged and written as:

$$a = \frac{1}{t_{ref} + \tau ln(1 + u_{th}/(I_o - u_{th}))} \tag{13.8}$$

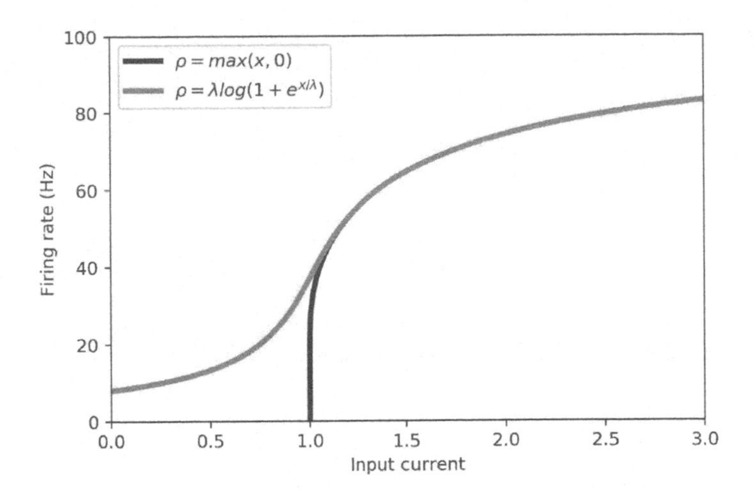

Figure 13.6 LIF tuning curve can be modified to provide a differentiable model. Parameters used: $R = 1{,}000, C = 5 \cdot 10^{-6}, t_{ref} = 10 \cdot 10^{-3}, u_{th} = 1, \lambda = 0.05$.

Note that R was embedded within the input term.

A rectified version of this equation would look like:

$$a = \frac{1}{t_{ref} + \tau ln(1 + u_{th}/\rho(I_o - u_{th}))} \tag{13.9}$$

where $\rho(x) = max(x, 0)$.

To provide a differentiable model, we can define ρ as a soft-max function:

$$\rho(x) = \lambda log(1 + e^{x/\lambda})$$

where λ is a smoothing factor. The standard and the smoothed tuning curves are shown in **Figure 13.6**.

Now we can convert a trained ANN to SNN solely by defining the neurons as soft LIF spiking neurons and temporally integrate their spikes to remove high-frequency variation. In such an implementation, the spiking neurons' differentiable approximations are used during training while the spiking neurons themselves are used during inference.

Figure 13.7 Examples of 16 images from the MNIST database.

13.3.1 MNIST classification with deep SNN

Note: This subsection is based on Nengo's documentation, also available at nengo.ai. Implementation details are also available on the book website.

We will showcase the classification of MNIST images. MNIST holds 70,000 examples of classified handwritten numbers (comprises a training set of 60,000 examples and a test set of 10,000 examples) [168]. An example of 16 images from the MNIST database is shown in **Figure 13.7**. Images in MNIST are centered in a 28×28 shape.

The model presented here is based on Nengo-DL [239], a framework with which a Nengo-based network can be simulated using some underlying computational framework, in this case, TensorFlow [1]. Google Brain developed Tensorflow as a software package with which large scale

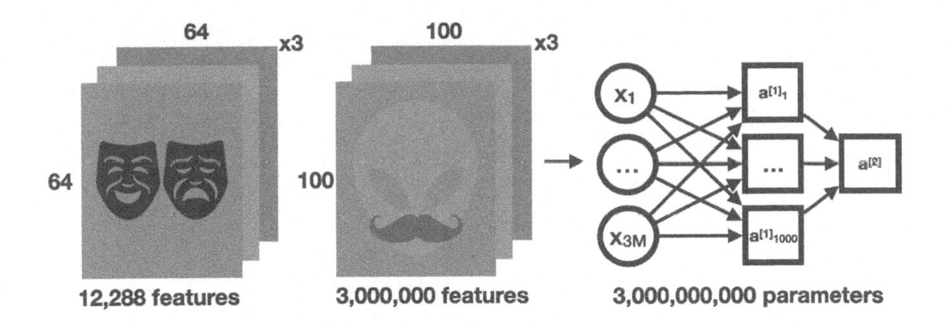

Figure 13.8 For image classification, a 64 × 64 × 3 image would have 12,288 features and a 100 × 100 × 3 image 3,000,000 features. A DNN with a single fully connected hidden layer comprises 1,000 neurons and entails more than 3 billion parameters.

ANNs can be defined and implemented across multiple computational frameworks (e.g., GPU, CPU, and cloud).

To perform MNIST classification, we will shortly describe CNNs.

13.3.1.1 Convolutional neural networks

In a traditional DNN for image classification, each pixel is a feature (input to the model). Therefore, a 64 × 64 color image would have 12,288 features, and a 100 × 100 color image would have 3,000,000 features. Building a DNN with a single fully connected hidden layer comprised of 1,000 neurons would entail more than 3 billion parameters (**Figure 13.8**). In such an enormous model, it would be hard to get enough data to prevent over-fitting and manage memory efficiently. As we will see, the convolutional approach to neural networks dramatically reduces the number of parameters.

In Eq. 12.15, we described convolution in continuous space. When we work with images which are discretized entities we can define convolution as a discrete operator. For a 1D signal, we will define A as a mask and I as a flattened 1D image, and formulate the convolution with:

$$A * I[x] = \int_{i=-\frac{m}{2}}^{\frac{m}{2}} A(i)I(x-i) \tag{13.10}$$

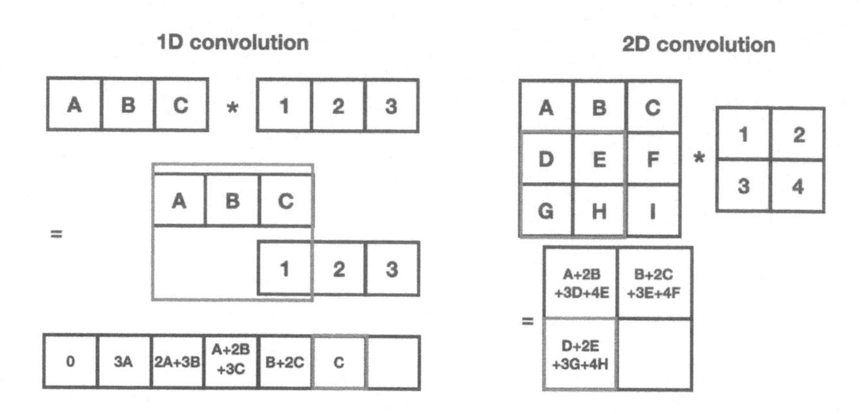

Figure 13.9 Illustration of 1D and 2D convolutions.

In 2D images, we will similarly define the 2D convolution operator:

$$A * I[x,y] = \int_{i=-\frac{m}{2}}^{\frac{m}{2}} \int_{j=-\frac{m}{2}}^{\frac{m}{2}} A(i,j)I(x-i,y-j) \tag{13.11}$$

Both 1D and 2D convolutions are illustrated in **Figure 13.9**.

We can move forward to define convolutions in 3D. Note that convolution can reduce the size of the resulting image and therefore, pixels at the image's edges contribute less to the output. Image padding can be applied to prevent image downsizing. For example, padding a 3 × 3 image with additional pixels (p = 1, value = 0) around the edges will result in a 5 × 5 image. When convolved with a 3 × 3 filter (or kernel), the result is a 3 × 3 image (**Figure 13.10**).

Another important modulation of convolution is striding where positions are skipped during convolution. For example, convolving a 7 × 7 image with a 3 × 3 kernel, and a stride = 2 results in a 3 × 3 image (**Figure 13.11**).

13.3.1.2 CNN architecture

In this example, we will define the following CNN architecture. A 28 × 28 image is introduced to the first convolutional layer, where it is convolved with 32 3 × 3 kernels, producing 32 26 × 26 images. Note that (1) image size is reduced due to unpadded convolution, and (2) by convolving the image with a few kernels, we have a 3D matrix as a result (here, a 26 × 26 × 32 matrix). This 26 × 26 × 32 matrix is introduced into a second convolutional layer where it is convolved with 64 3 × 3 ×

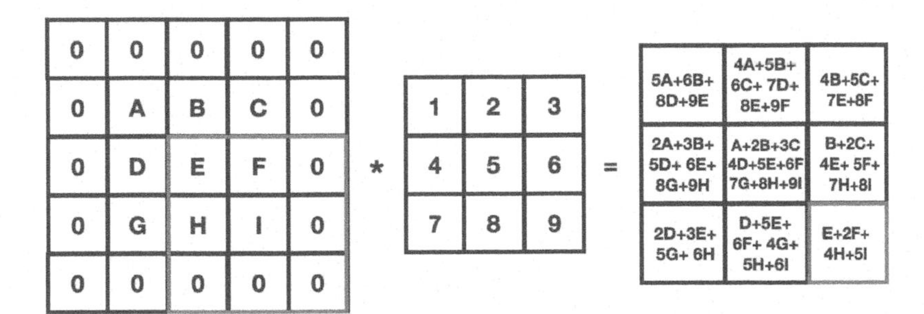

Figure 13.10 By padding an image with additional pixels layers, image downsizing during convolution is prevented. Padding a 3 × 3 image (p = 1, value = 0) creates a 5 × 5 image. When convolved with a 3 × 3 kernel, the result is a 3 × 3.

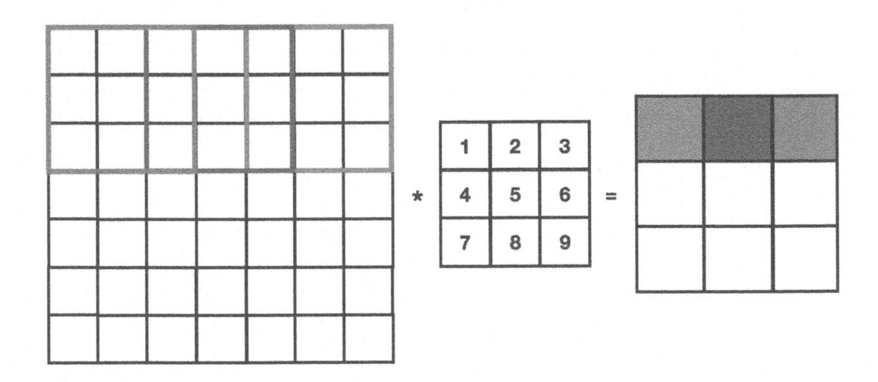

Figure 13.11 In a strided convolution, we jump over positions during convolution. Convolving a 7 × 7 image with a 3 × 3 kernel and stride = 2 results in a 3 × 3 image.

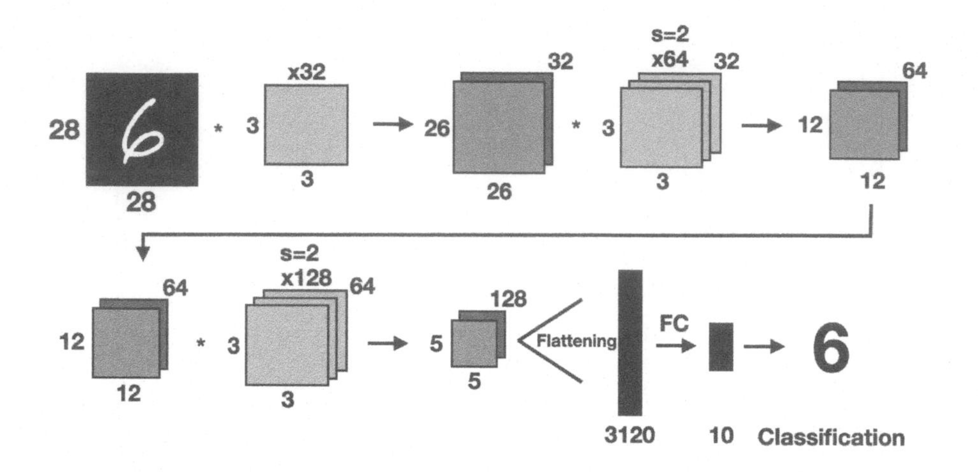

Figure 13.12 CNN architecture for MNIST classification, comprising of three convolutional layers and one fully connected layer. Orange squares represent kernels, pink-purple squares represent intermediate matrices, s is the striding parameter and FC stands for fully connected.

32 kernels. In this layer, a striding of 2 is used, generating 64 12×12 images. The resulting $12 \times 12 \times 64$ matrix is now introduced into a third convolutional layer where it is convolved with 128 $3 \times 3 \times 64$ kernels, with a striding of 2, generating a $5 \times 5 \times 128$ matrix. Finally, all 3,200 values are fully connected to a hidden layer comprised of 10 neurons. The value encoded by each neuron should correspond to the number of classification classes, a number ranging from 0 to 9. CNN architecture is summarized in **Figure 13.12**. The network was defined with LIF neurons with a maximal firing rate of 100 Hz and a zero intercept.

13.3.1.3 Results

Note that training is held using a rate-based approximation, with soft LIF neurons, as was described above. Therefore, during training, we do not need a notion of time. The simulation is, therefore, configured to run in a single time step. However, we use spiking neuron models for inference for which spikes are temporally integrated across multiple time-steps (here, we defined 30 time-steps). Another implementation detail is the quantification of the network performance. Since it takes time for spike data to integrate, we evaluate the network's output by considering the final time-step. Following training, we measure an

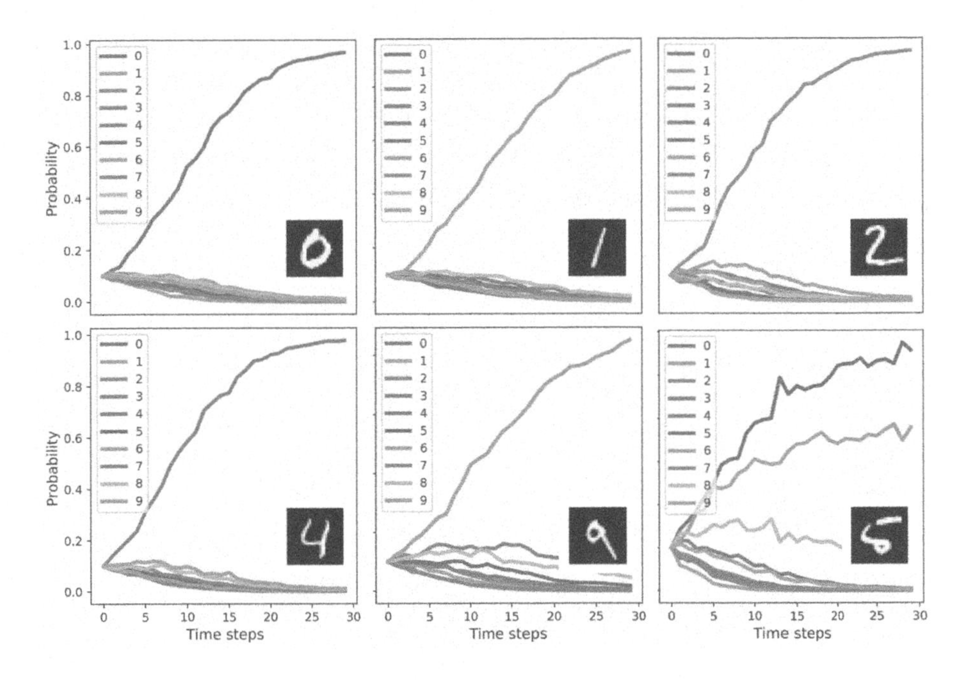

Figure 13.13 Examples of five correctly classified images and one mis-classified image, generated by the spiking CNN shown in Figure 13.13.

accuracy of 98.7%. Examples of five correctly classified examples and one mis-classified example are shown in **Figure 13.13**.

13.4 GLOSSARY

Prescribed error sensitivity: A biologically plausible NEF-based supervised learning rule which modifies connection weights between populations of neurons to minimize an external error signal

13.5 FURTHER READING

- **Section 13.1**
  - Reviews of neuromorphic learning algorithm are given in [248] and [272]. A timing-tailored learning review is given in [196].
- **Section 13.2**
  - The PES rule is described along with Hebbian-based learning in [31] and analytically evaluated in [294].

- **Section 13.3**
  - MNIST description, as well as references to prominent classification algorithms are available on yann.lecun.com
  - NengoDL documentation is available on nengo.ai/nengo-dl and its utilization for MNIST classification is available on nengo.ai/nengo-dl

Bibliography

[1] Martín Abadi et al. "Tensorflow: A system for large-scale machine learning". In: *12th Symposium on Operating Systems Design and Implementation (OSDI 16)*. 2016, pp. 265–283.

[2] Larry F Abbott. "Lapicque's introduction of the integrate-and-fire model neuron (1907)". In: *Brain Research Bulletin* 50.5-6 (1999), pp. 303–304.

[3] Aanen Abusland, Tor S Lande, and M Hovin. "A VLSI communication architecture for stochastically pulse-encoded analog signals". In: *1996 IEEE International Symposium on Circuits and Systems. Circuits and Systems Connecting the World. ISCAS 96.* Vol. 3. IEEE. 1996, pp. 401–404.

[4] David H Ackley, Geoffrey E Hinton, and Terrence J Sejnowski. "A learning algorithm for Boltzmann machines". In: *Cognitive Science* 9.1 (1985), pp. 147–169.

[5] Andrew Adamatzky. *Game of life cellular automata.* Vol. 1. Springer, 2010.

[6] Andrew Adamatzky and Leon Chua. *Memristor networks.* Springer Science & Business Media, 2013.

[7] Sally Adee. "Cat-brain fever". In: *IEEE Spectrum* 47.1 (2010), pp. 16–17.

[8] Matt Ainsworth et al. "Rates and rhythms: a synergistic view of frequency and temporal coding in neuronal networks". In: *Neuron* 75.4 (2012), pp. 572–583.

[9] Sacha J van Albada et al. "Performance comparison of the digital neuromorphic hardware SpiNNaker and the neural network simulation software NEST for a full-scale cortical microcircuit model". In: *Frontiers in Neuroscience* 12 (2018), p. 291.

[10] Mara Almog and Alon Korngreen. "Is realistic neuronal modeling realistic?" In: *Journal of Neurophysiology* 116.5 (2016), pp. 2180–2209.

[11] Arnon Amir et al. "Cognitive computing programming paradigm: a Corelet Language for composing networks of neurosynaptic cores". In: *The 2013 International Joint Conference on Neural Networks (IJCNN)*. IEEE. 2013, pp. 1–10.

[12] Daniel J Amit and Daniel J Amit. *Modeling brain function: The world of attractor neural networks*. Cambridge University Press, 1992.

[13] Oren Amsalem et al. "An efficient analytical reduction of detailed nonlinear neuron models". In: *Nature Communications* 11.1 (2020), pp. 1–13.

[14] Rajagopal Ananthanarayanan and Dharmendra S Modha. "Anatomy of a cortical simulator". In: *SC'07: Proceedings of the 2007 ACM/IEEE Conference on Supercomputing*. IEEE. 2007, pp. 1–12.

[15] Rajagopal Ananthanarayanan et al. "The cat is out of the bag: cortical simulations with 109 neurons, 1013 synapses". In: *Proceedings of the Conference on High Performance Computing Networking, Storage and Analysis*. 2009, pp. 1–12.

[16] Leon Anavy et al. "Data storage in DNA with fewer synthesis cycles using composite DNA letters". In: *Nature Biotechnology* 37.10 (2019), pp. 1229–1236.

[17] David J Anderson. "Circuit modules linking internal states and social behaviour in flies and mice". In: *Nature Reviews Neuroscience* 17.11 (2016), p. 692.

[18] William S Anderson and Gabriel Kreiman. "Neuroscience: What we cannot model, we do not understand". In: *Current Biology* 21.3 (2011), R123–R125.

[19] Lea Ankri et al. "Antagonistic center-surround mechanisms for direction selectivity in the retina". In: *Cell reports* 31.5 (2020), p. 107608.

[20] Ismail Emre Araci and Stephen R Quake. "Microfluidic very large scale integration (mVLSI) with integrated micromechanical valves". In: *Lab on a Chip* 12.16 (2012), pp. 2803–2806.

[21] John V Arthur et al. "Building block of a programmable neuromorphic substrate: A digital neurosynaptic core". In: *The 2012 International Joint Conference on Neural Networks (IJCNN)*. IEEE. 2012, pp. 1–8.

[22] Frank Arute et al. "Quantum supremacy using a programmable superconducting processor". In: *Nature* 574.7779 (2019), pp. 505–510.

[23] John Backus. "Can programming be liberated from the von Neumann style? A functional style and its algebra of programs". In: *Communications of the ACM* 21.8 (1978), pp. 613–641.

[24] Sunny Bains. "The business of building brains". In: *Nature Electronics* 3.7 (2020), pp. 348–351.

[25] Y. Barkan and H. Spitzer. "The Color Dove Illusion:Chromatic Filling-In Effect Following a Spatial-Temporal Edge, in A. G. Shapiro and D. Todorovic (eds.)" In: *The Oxford Compendium of Visual Illusions* (2017).

[26] Horace B Barlow. "'Single units and sensation: A neuron doctrine for perceptual psychology?': Author's update". In: *Perception* (2009).

[27] Horace B Barlow. "Single units and sensation: a neuron doctrine for perceptual psychology?" In: *Perception* 1.4 (1972), pp. 371–394.

[28] Natali Barros Zulaica et al. "Estimating the readily-releasable vesicle pool size at synaptic connections in a neocortical microcircuit". In: *Frontiers in Synaptic Neuroscience* 11 (2019), p. 29.

[29] Chiara Bartolozzi and Giacomo Indiveri. "Synaptic dynamics in analog VLSI". In: *Neural Computation* 19.10 (2007), pp. 2581–2603.

[30] Mark Bear, Barry Connors, and Michael A Paradiso. *Neuroscience: Exploring the Brain*. Jones & Bartlett Learning, LLC, 2020.

[31] Trevor Bekolay, Carter Kolbeck, and Chris Eliasmith. "Simultaneous unsupervised and supervised learning of cognitive functions in biologically plausible spiking neural networks". In: *Proceedings of the Annual Meeting of the Cognitive Science Society*. Vol. 35. 35. 2013.

[32] Trevor Bekolay et al. "Nengo: a Python tool for building large-scale functional brain models". In: *Frontiers in Neuroinformatics* 7 (2014), p. 48.

[33] Ben Varkey Benjamin et al. "Neurogrid: A mixed-analog-digital multichip system for large-scale neural simulations". In: *Proceedings of the IEEE* 102.5 (2014), pp. 699–716.

[34] Charles H Bennett. "Universal computation and physical dynamics". In: *Physica D: Nonlinear phenomena* 86.1-2 (1995), pp. 268–273.

[35] Maxwell R Bennett et al. *Neuroscience and philosophy: Brain, mind, and language.* Columbia University Press, 2007.

[36] Elie L Bienenstock, Leon N Cooper, and Paul W Munro. "Theory for the development of neuron selectivity: orientation specificity and binocular interaction in visual cortex". In: *Journal of Neuroscience* 2.1 (1982), pp. 32–48.

[37] Zhenshan Bing et al. "A survey of robotics control based on learning-inspired spiking neural networks". In: *Frontiers in Neurorobotics* 12 (2018), p. 35.

[38] Kwabena A Boahen. "A burst-mode word-serial address-event Link-III: Analysis and test results". In: *IEEE Transactions on Circuits and Systems I: Regular Papers* 51.7 (2004), pp. 1292–1300.

[39] Kwabena A Boahen. "A burst-mode word-serial address-event link-I: Transmitter design". In: *IEEE Transactions on Circuits and Systems I: Regular Papers* 51.7 (2004), pp. 1269–1280.

[40] Kwabena A Boahen. "A burst-mode word-serial address-event link-II: Receiver design". In: *IEEE Transactions on Circuits and Systems I: Regular Papers* 51.7 (2004), pp. 1281–1291.

[41] Kwabena A Boahen. "Point-to-point connectivity between neuromorphic chips using address events". In: *IEEE Transactions on Circuits and Systems II: Analog and Digital Signal Processing* 47.5 (2000), pp. 416–434.

[42] Kwabena Boahen. "A neuromorph's prospectus". In: *Computing in Science & Engineering* 19.2 (2017), pp. 14–28.

[43] Rafal Bogacz. "Optimal decision-making theories: linking neurobiology with behaviour". In: *Trends in Cognitive Sciences* 11.3 (2007), pp. 118–125.

[44] Sander M Bohte, Joost N Kok, and Han La Poutre. "Error-backpropagation in temporally encoded networks of spiking neurons". In: *Neurocomputing* 48.1-4 (2002), pp. 17–37.

[45] Sander M Bohte, Joost N Kok, and Johannes A La Poutré. "SpikeProp: backpropagation for networks of spiking neurons". In: *ESANN*. Vol. 48. 2000, pp. 17–37.

[46] Nick Bostrom and Anders Sandberg. "Whole brain emulation: a roadmap". In: *Lanc Univ Accessed January* 21.2008 (2008), p. 2015.

[47] Nicolas Brunel and Mark CW Van Rossum. "Lapicque's 1907 paper: from frogs to integrate-and-fire". In: *Biological Cybernetics* 97.5-6 (2007), pp. 337–339.

[48] Anthony N Burkitt. "A review of the integrate-and-fire neuron model: II. Inhomogeneous synaptic input and network properties". In: *Biological Cybernetics* 95.2 (2006), pp. 97–112.

[49] Natalia Caporale and Yang Dan. "Spike timing–dependent plasticity: a Hebbian learning rule". In: *Annual Review of Neuroscience* 31 (2008), pp. 25–46.

[50] Francesco Caravelli and Juan Pablo Carbajal. "Memristors for the curious outsiders". In: *Technologies* 6.4 (2018), p. 118.

[51] Andrew S Cassidy et al. "Cognitive computing building block: A versatile and efficient digital neuron model for neurosynaptic cores". In: *The 2013 International Joint Conference on Neural Networks (IJCNN)*. IEEE. 2013, pp. 1–10.

[52] William A Catterall et al. "The Hodgkin-Huxley heritage: from channels to circuits". In: *Journal of Neuroscience* 32.41 (2012), pp. 14064–14073.

[53] Gastone G Celesia. "Alcmaeon of Croton's observations on health, brain, mind, and soul". In: *Journal of the History of the Neurosciences* 21.4 (2012), pp. 409–426.

[54] Enea Ceolini et al. "Hand-gesture recognition based on EMG and event-based camera sensor fusion: a benchmark in neuromorphic computing". In: *Frontiers in Neuroscience* 14 (2020).

[55] Lih Feng Cheow, Levent Yobas, and Dim-Lee Kwong. "Digital microfluidics: Droplet based logic gates". In: *Applied Physics Letters* 90.5 (2007), p. 054107.

[56] Yoonsuck Choe. “Hebbian Learning”. In: *Encyclopedia of Computational Neuroscience*. Ed. by Dieter Jaeger and Ranu Jung. New York, NY: Springer New York, 2013, pp. 1–5.

[57] Feng-Xuan Choo. “Spaun 2.0: Extending the World’s Largest Functional Brain Model”. 2018.

[58] Leon Chua. “Memristor-the missing circuit element”. In: *IEEE Transactions on Circuit Theory* 18.5 (1971), pp. 507–519.

[59] Jay S Coggan et al. “A process for digitizing and simulating biologically realistic oligocellular networks demonstrated for the neuro-glio-vascular ensemble”. In: *Frontiers in Neuroscience* 12 (2018), p. 664.

[60] Hadar Cohen-Duwek and Elishai Ezra Tsur. “Biologically Plausible Spiking Neural Networks for Perceptual Filling-In”. In: 2021.

[61] Hadar Cohen-Duwek and Hedva Spitzer. *A Compound Computational Model for Filling-in Processes Triggered by Edges: Watercolor Illusions.* Frontiers in neuroscience. 2019; 13:225. Published online 2019 Mar 22. doi: 10.3389/fnins.2019.00225.

[62] Leon N Cooper and Mark F Bear. “The BCM theory of synapse modification at 30: interaction of theory with experiment”. In: *Nature Reviews Neuroscience* 13.11 (2012), pp. 798–810.

[63] Francesco Cremonesi et al. “Analytic performance modeling and analysis of detailed neuron simulations”. In: *The International Journal of High Performance Computing Applications* 34.4 (2020), pp. 428–449.

[64] Travis Crncich-DeWolf et al. *Methods and systems for nonlinear adaptive control and filtering.* US Patent 10,481,565. 2019.

[65] Sharon Crook et al. “MorphML: Level 1 of the NeuroML standards for neuronal morphology data and model specification”. In: *Neuroinformatics* 5.2 (2007), pp. 96–104.

[66] Kael Dai et al. “The SONATA data format for efficient description of large-scale network models”. In: *PLoS computational biology* 16.2 (2020), e1007696.

[67] Henry Dale. “Pharmacology and nerve-endings”. In: *The Journal of Nervous and Mental Disease* 82.4 (1935), p. 457.

[68] Tanguy Damart, Werner Van Geit, and Henry Markram. "Data driven building of realistic neuron model using IBEA and CMA evolution strategies". In: *Proceedings of the 2020 Genetic and Evolutionary Computation Conference Companion.* 2020, pp. 35–36.

[69] Mike Davies et al. "Loihi: A neuromorphic manycore processor with on-chip learning". In: *IEEE Micro* 38.1 (2018), pp. 82–99.

[70] Andrew P Davison et al. "PyNN: a common interface for neuronal network simulators". In: *Frontiers in Neuroinformatics* 2 (2009), p. 11.

[71] Benedetto De Martino et al. "Frames, biases, and rational decision-making in the human brain". In: *Science* 313.5787 (2006), pp. 684–687.

[72] Michael V DeBole et al. "TrueNorth: Accelerating from zero to 64 million neurons in 10 years". In: *Computer* 52.5 (2019), pp. 20–29.

[73] Travis DeWolf, Pawel Jaworski, and Chris Eliasmith. "Nengo and low-power AI hardware for robust, embedded neurorobotics". In: *arXiv preprint arXiv:2007.10227* (2020).

[74] Travis DeWolf et al. "A spiking neural model of adaptive arm control". In: *Proceedings of the Royal Society B: Biological Sciences* 283.1843 (2016), p. 20162134.

[75] Peter U Diehl and Matthew Cook. "Efficient implementation of STDP rules on SpiNNaker neuromorphic hardware". In: *2014 International Joint Conference on Neural Networks (IJCNN).* IEEE. 2014, pp. 4288–4295.

[76] Peter U Diehl and Matthew Cook. "Unsupervised learning of digit recognition using spike-timing-dependent plasticity". In: *Frontiers in Computational Neuroscience* 9 (2015), p. 99.

[77] Peter U Diehl et al. "Truehappiness: Neuromorphic emotion recognition on TrueNorth". In: *2016 International Joint Conference on Neural Networks (IJCNN).* IEEE. 2016, pp. 4278–4285.

[78] Elisa Donati et al. "Processing EMG signals using reservoir computing on an event-based neuromorphic system". In: *2018 IEEE Biomedical Circuits and Systems Conference (BioCAS).* IEEE. 2018, pp. 1–4.

[79] Rodney Douglas, Misha Mahowald, and Carver Mead. "Neuromorphic analogue VLSI". In: *Annual Review of Neuroscience* 18.1 (1995), pp. 255–281.

[80] Zidong Du et al. "Neuromorphic accelerators: A comparison between neuroscience and machine-learning approaches". In: *2015 48th Annual IEEE/ACM International Symposium on Microarchitecture (MICRO).* IEEE. 2015, pp. 494–507.

[81] Gaute T Einevoll et al. "The scientific case for brain simulations". In: *Neuron* 102.4 (2019), pp. 735–744.

[82] Miriam Elbaz, Rachel Buterman, and Elishai Ezra Tsur. "NeuroConstruct-based implementation of structured-light stimulated retinal circuitry". In: *BMC Neuroscience* 21.1 (2020), pp. 1–9.

[83] Chris Eliasmith. *How to build a brain: A neural architecture for biological cognition.* Oxford University Press, 2013.

[84] Chris Eliasmith and Charles H Anderson. *Neural engineering: Computation, representation, and dynamics in neurobiological systems.* MIT Press, 2003.

[85] Chris Eliasmith, Jan Gosmann, and Xuan Choo. "BioSpaun: A large-scale behaving brain model with complex neurons". In: *arXiv preprint arXiv:1602.05220* (2016).

[86] Chris Eliasmith and Oliver Trujillo. "The use and abuse of large-scale brain models". In: *Current Opinion in Neurobiology* 25 (2014), pp. 1–6.

[87] Chris Eliasmith et al. "A large-scale model of the functioning brain". In: *Science* 338.6111 (2012), pp. 1202–1205.

[88] Roger M Enoka and Jacques Duchateau. "Rate coding and the control of muscle force". In: *Cold Spring Harbor Perspectives in Medicine* 7.10 (2017), a029702.

[89] Csaba Erö et al. "A cell atlas for the mouse brain". In: *Frontiers in Neuroinformatics* 12 (2018), p. 84.

[90] Steve K Esser et al. "Cognitive computing systems: Algorithms and applications for networks of neurosynaptic cores". In: *The 2013 International Joint Conference on Neural Networks (IJCNN).* IEEE. 2013, pp. 1–10.

[91] Xue Fan and Henry Markram. "A brief history of simulation neuroscience". In: *Frontiers in Neuroinformatics* 13 (2019), p. 32.

[92] Richard P Feynman. "Simulating physics with computers". In: *International Journal of Theoretical Physics* 21.6/7 (1982).

[93] Luis M Fidalgo and Sebastian J Maerkl. "A software-programmable microfluidic device for automated biology". In: *Lab on a Chip* 11.9 (2011), pp. 1612–1619.

[94] Greg D Field and EJ Chichilnisky. "Information processing in the primate retina: circuitry and coding". In: *Annu. Rev. Neurosci.* 30 (2007), pp. 1–30.

[95] James W Fransen and Bart G Borghuis. "Temporally diverse excitation generates direction-selective responses in ON-and OFF-type retinal starburst amacrine cells". In: *Cell Reports* 18.6 (2017), pp. 1356–1365.

[96] Steve B Furber et al. "The SpiNNaker project". In: *Proceedings of the IEEE* 102.5 (2014), pp. 652–665.

[97] Hagar Gelbard-Sagiv et al. "Internally generated reactivation of single neurons in human hippocampus during free recall". In: *Science* 322.5898 (2008), pp. 96–101.

[98] J Georgiou and AG Andreou. "High-speed, address-encoding arbiter architecture". In: *Electronics Letters* 42.3 (2006), p. 1.

[99] Wulfram Gerstner et al. "Neural codes: firing rates and beyond". In: *Proceedings of the National Academy of Sciences* 94.24 (1997), pp. 12740–12741.

[100] Wulfram Gerstner et al. *Neuronal dynamics: From single neurons to networks and models of cognition.* Cambridge University Press, 2014.

[101] Charles D Gilbert and Wu Li. "Top-down influences on visual processing". In: *Nature Reviews Neuroscience* 14.5 (2013), pp. 350–363.

[102] Jacob R Glaser and Edmund M Glaser. "Neuron imaging with Neurolucida—a PC-based system for image combining microscopy". In: *Computerized Medical Imaging and Graphics* 14.5 (1990), pp. 307–317.

[103] Padraig Gleeson, Volker Steuber, and R Angus Silver. "Neuro-Construct: a tool for modeling networks of neurons in 3D space". In: *Neuron* 54.2 (2007), pp. 219–235.

[104] Padraig Gleeson et al. "NeuroML: a language for describing data driven models of neurons and networks with a high degree of biological detail". In: *PLoS Comput Biol* 6.6 (2010), e1000815.

[105] Padraig Gleeson et al. "Open source brain: a collaborative resource for visualizing, analyzing, simulating, and developing standardized models of neurons and circuits". In: *Neuron* 103.3 (2019), pp. 395–411.

[106] Dan FM Goodman and Romain Brette. "The brian simulator". In: *Frontiers in Neuroscience* 3 (2009), p. 26.

[107] Jan Gosmann and Chris Eliasmith. "Automatic optimization of the computation graph in the Nengo neural network simulator". In: *Frontiers in Neuroinformatics* 11 (2017), p. 33.

[108] Alex Graves, Greg Wayne, and Ivo Danihelka. "Neural Turing Machines". In: *arXiv preprint arXiv:1410.5401* (2014).

[109] Matthew J Greene, Jinseop S Kim, H Sebastian Seung, et al. "Analogous convergence of sustained and transient inputs in parallel on and off pathways for retinal motion computation". In: *Cell Reports* 14.8 (2016), pp. 1892–1900.

[110] Charles G Gross. "Genealogy of the 'grandmother cell'". In: *The Neuroscientist* 8.5 (2002), pp. 512–518.

[111] Laszlo Gyongyosi and Sandor Imre. "A survey on quantum computing technology". In: *Computer Science Review* 31 (2019), pp. 51–71.

[112] Avi Hazan and Elishai Ezra Tsur. "Neuromorphic Analog Implementation of Neural Engineering Framework-Inspired Spiking Neuron for High-Dimensional Representation". In: *Frontiers in Neuroscience* 15 (2021), p. 109.

[113] Donald Olding Hebb. *The organization of behavior: A neuropsychological theory.* Psychology Press, 2005.

[114] Ori Heimlich and Elishai Ezra Tsur. "OpenVX-Based Python Framework for Real-time Cross-Platform Acceleration of embedded Computer Vision Applications". In: *Frontiers in ICT* 3 (2016), p. 28.

[115] Moritz Helmstaedter et al. "Reconstruction of an average cortical column in silico". In: *Brain Research Reviews* 55.2 (2007), pp. 193–203.

[116] Michael Hennecke et al. "Measuring power consumption on IBM Blue Gene/P". In: *Computer Science-Research and Development* 27.4 (2012), pp. 329–336.

[117] John L Hennessy and David A Patterson. *Computer Architecture: A Quantitative Approach*. Elsevier, 2011.

[118] Frank Heppner and Ulf Grenander. "A stochastic nonlinear model for coordinated bird flocks". In: *The Ubiquity of Chaos* 233 (1990), p. 238.

[119] Suzana Herculano-Houzel. "The human brain in numbers: a linearly scaled-up primate brain". In: *Frontiers in Human Neuroscience* 3 (2009), p. 31.

[120] Andreas VM Herz et al. "Modeling single-neuron dynamics and computations: a balance of detail and abstraction". In: *Science* 314.5796 (2006), pp. 80–85.

[121] Rudiger von der Heydt, Howard S Friedman, and Hong Zhou. "Searching for the neural mechanisms of color filling-in". In: *Filling-in: From Perceptual Completion to Cortical Reorganization* (2003), pp. 106–127.

[122] W Daniel Hillis. "Intelligence as an emergent behavior; or, the songs of Eden". In: *Daedalus* (1988), pp. 175–189.

[123] Michael L Hines and Nicholas T Carnevale. "The NEURON simulation environment". In: *Neural Computation* 9.6 (1997), pp. 1179–1209.

[124] Michael L Hines, Hubert Eichner, and Felix Schürmann. "Neuron splitting in compute-bound parallel network simulations enables runtime scaling with twice as many processors". In: *Journal of Computational Neuroscience* 25.1 (2008), pp. 203–210.

[125] Michael Hines, Andrew P Davison, and Eilif Muller. "NEURON and Python". In: *Frontiers in Neuroinformatics* 3 (2009), p. 1.

[126] Alan L Hodgkin and Andrew F Huxley. "A quantitative description of membrane current and its application to conduction and excitation in nerve". In: *The Journal of Physiology* 117.4 (1952), p. 500.

[127] William R. Holmes. “Cable Equation”. In: *Encyclopedia of Computational Neuroscience.* Ed. by Dieter Jaeger and Ranu Jung. New York, NY: Springer New York, 2013, pp. 1–13.

[128] Jong Wook Hong and Stephen R Quake. “Integrated nanoliter systems”. In: *Nature Biotechnology* 21.10 (2003), pp. 1179–1183.

[129] Sang Wook Hong and Frank Tong. “Neural representation of form-contingent color filling-in in the early visual cortex”. In: *Journal of Vision* 17.13 (2017), pp. 10–10.

[130] John J Hopfield. “Neural networks and physical systems with emergent collective computational abilities”. In: *Proceedings of the National Academy of Sciences* 79.8 (1982), pp. 2554–2558.

[131] Scarlett R Howard et al. “Numerical ordering of zero in honey bees”. In: *Science* 360.6393 (2018), pp. 1124–1126.

[132] David H Hubel and Torsten N Wiesel. “Receptive fields of single neurones in the cat’s striate cortex”. In: *The Journal of Physiology* 148.3 (1959), p. 574.

[133] Eric Hunsberger and Chris Eliasmith. “Spiking deep networks with LIF neurons”. In: *arXiv preprint arXiv:1510.08829* (2015).

[134] Eric Hunsberger and Chris Eliasmith. “Training spiking deep networks for neuromorphic hardware”. In: *arXiv preprint arXiv:1611.05141* (2016).

[135] Antje Ihlefeld, Nima Alamatsaz, and Robert M Shapley. “Population rate-coding predicts correctly that human sound localization depends on sound intensity”. In: *Elife* 8 (2019), e47027.

[136] Nabil Imam and Rajit Manohar. “Address-event communication using token-ring mutual exclusion”. In: *2011 17th IEEE International Symposium on Asynchronous Circuits and Systems.* IEEE. 2011, pp. 99–108.

[137] Giacomo Indiveri and Rodney Douglas. “Neuromorphic vision sensors”. In: *Science* 288.5469 (2000), pp. 1189–1190.

[138] Giacomo Indiveri et al. “Neuromorphic silicon neuron circuits”. In: *Frontiers in Neuroscience* 5 (2011), p. 73.

[139] Thomas R Insel, Story C Landis, and Francis S Collins. “The NIH brain initiative”. In: *Science* 340.6133 (2013), pp. 687–688.

[140] Takahiro Ishikawa et al. "The cerebro-cerebellum: Could it be loci of forward models?" In: *Neuroscience Research* 104 (2016), pp. 72–79.

[141] Eugene M Izhikevich. "Simple model of spiking neurons". In: *IEEE Transactions on Neural Networks* 14.6 (2003), pp. 1569–1572.

[142] Eugene M Izhikevich. "Which model to use for cortical spiking neurons?" In: *IEEE Transactions on Neural Networks* 15.5 (2004), pp. 1063–1070.

[143] Eugene M Izhikevich and Niraj S Desai. "Relating stdp to bcm". In: *Neural Computation* 15.7 (2003), pp. 1511–1523.

[144] Eugene M Izhikevich and Gerald M Edelman. "Large-scale model of mammalian thalamocortical systems". In: *Proceedings of the National Academy of Sciences* 105.9 (2008), pp. 3593–3598.

[145] Inez Jabalpurwala. "Brain Canada: one brain one community". In: *Neuron* 92.3 (2016), pp. 601–606.

[146] Ferris Jabr. "Does thinking really hard burn more calories?" In: *Scientific American* 18 (2012).

[147] Anil K Jain, Jianchang Mao, and K Moidin Mohiuddin. "Artificial neural networks: A tutorial". In: *Computer* 29.3 (1996), pp. 31–44.

[148] Sung Hyun Jo et al. "Nanoscale memristor device as synapse in neuromorphic systems". In: *Nano Letters* 10.4 (2010), pp. 1297–1301.

[149] Judson P Jones and Larry A Palmer. "An evaluation of the two-dimensional Gabor filter model of simple receptive fields in cat striate cortex". In: *Journal of Neurophysiology* 58.6 (1987), pp. 1233–1258.

[150] Lila Kari and Grzegorz Rozenberg. "The many facets of natural computing". In: *Communications of the ACM* 51.10 (2008), pp. 72–83.

[151] Y Katsuki. "Neural mechanism of auditory sensation in cats". In: *Sensory Communication* 561.584 (1961), pp. 561–584.

[152] Evgeny Katz. "Biocomputing—tools, aims, perspectives". In: *Current Opinion in Biotechnology* 34 (2015), pp. 202–208.

[153] Patrick W Keeley et al. "Dendritic spread and functional coverage of starburst amacrine cells". In: *Journal of Comparative Neurology* 505.5 (2007), pp. 539–546.

[154] K. F. Kelly and C. C. M. Mody. "The booms and busts of molecular electronics". In: *IEEE Spectrum* 52.10 (2015), pp. 52–60.

[155] Robert W Keyes. "The future of the transistor". In: *Scientific American* 8.1 (1997), pp. 46–52.

[156] Saeed Reza Kheradpisheh et al. "STDP-based spiking deep convolutional neural networks for object recognition". In: *Neural Networks* 99 (2018), pp. 56–67.

[157] Jinseop S Kim et al. "Space–time wiring specificity supports direction selectivity in the retina". In: *Nature* 509.7500 (2014), pp. 331–336.

[158] James C Knight and Thomas Nowotny. "GPUs outperform current HPC and neuromorphic solutions in terms of speed and energy when simulating a highly-connected cortical model". In: *Frontiers in Neuroscience* 12 (2018), p. 941.

[159] Peter Kogge. "The tops in flops". In: *IEEE Spectrum* 48.2 (2011), pp. 48–54.

[160] Hidehiko Komatsu. "The neural mechanisms of perceptual filling-in". In: *Nature Reviews Neuroscience* 7.3 (2006), pp. 220–231.

[161] Pramod Kumbhar et al. "CoreNEURON: an optimized compute engine for the NEURON simulator". In: *Frontiers in Neuroinformatics* 13 (2019), p. 63.

[162] Johan Kwisthout and Nils Donselaar. "On the Computational Power and Complexity of Spiking Neural Networks". In: *Proceedings of the Neuro-Inspired Computational Elements Workshop*. NICE '20. Heidelberg, Germany: Association for Computing Machinery, 2020.

[163] Madeline A Lancaster et al. "Cerebral organoids model human brain development and microcephaly". In: *Nature* 501.7467 (2013), pp. 373–379.

[164] Madeline A Lancaster et al. "Guided self-organization and cortical plate formation in human brain organoids". In: *Nature Biotechnology* 35.7 (2017), pp. 659–666.

[165] Louis Lapicque. "Recherches quantitatives sur l'excitation electrique des nerfs traitee comme une polarization". In: *Journal de Physiologie et de Pathologie Generalej* 9 (1907), pp. 620–635.

[166] Andrea Lavazza and Marcello Massimini. "Cerebral Organoids: Ethical Issues and Consciousness Assessment". In: *Journal of Medical Ethics* 44.9 (2018), pp. 606–610.

[167] John Lazzaro and John Wawrzynek. "Low-power silicon neurons, axons and synapses". In: *Silicon Implementation of Pulse Coded Neural Networks*. Springer, 1994, pp. 153–164.

[168] Yann LeCun et al. "Gradient-based learning applied to document recognition". In: *Proceedings of the IEEE* 86.11 (1998), pp. 2278–2324.

[169] Alex Lenz et al. "An adaptive gaze stabilization controller inspired by the vestibulo-ocular reflex". In: *Bioinspiration & Biomimetics* 3.3 (2008), p. 035001.

[170] Arnon Levy. "What was Hodgkin and Huxley's achievement?" In: *The British Journal for the Philosophy of Science* 65.3 (2014), pp. 469–492.

[171] Michael S Lewicki and Terrence J Sejnowski. "Learning overcomplete representations". In: *Neural Computation* 12.2 (2000), pp. 337–365.

[172] Qianli Liao, Joel Z Leibo, and Tomaso Poggio. "How important is weight symmetry in backpropagation?" In: *arXiv preprint arXiv:1510.05067* (2015).

[173] Jens Lienig and Juergen Scheible. *Fundamentals of Layout Design for Electronic Circuits*. Springer, 2020.

[174] Rob van Lier, Mark Vergeer, and Stuart Anstis. "Filling-in afterimage colors between the lines". In: *Current Biology* 19.8 (2009), R323–R324.

[175] Timothy P Lillicrap et al. "Backpropagation and the brain". In: *Nature Reviews Neuroscience* (2020), pp. 1–12.

[176] Chit-Kwan Lin et al. "Programming spiking neural networks on intel's Loihi". In: *Computer* 51.3 (2018), pp. 52–61.

[177] Abninder Litt et al. "Is the brain a quantum computer?" In: *Cognitive Science* 30.3 (2006), pp. 593–603.

[178] Qinghua Liu et al. "DNA computing on surfaces". In: *Nature* 403.6766 (2000), pp. 175–179.

[179] Shih-Chii Liu et al. *Event-based Neuromorphic Systems*. John Wiley & Sons, 2014.

[180] Rodolfo R Llinás. "The contribution of Santiago Ramon y Cajal to functional neuroscience". In: *Nature Reviews Neuroscience* 4.1 (2003), pp. 77–80.

[181] Maximilian PR Löhr, Christian Jarvers, and Heiko Neumann. "Complex neuron dynamics on the IBM TrueNorth neurosynaptic system". In: *2020 2nd IEEE International Conference on Artificial Intelligence Circuits and Systems (AICAS)*. IEEE. 2020, pp. 113–117.

[182] Wei Lu and Charles M Lieber. "Nanoelectronics from the bottom up". In: *Nanoscience And Technology: A Collection of Reviews from Nature Journals*. World Scientific, 2010, pp. 137–146.

[183] Richard F Lyon and Carver Mead. "An analog electronic cochlea". In: *IEEE Transactions on Acoustics, Speech, and Signal Processing* 36.7 (1988), pp. 1119–1134.

[184] Wolfgang Maass. "Lower bounds for the computational power of networks of spiking neurons". In: *Neural Computation* 8.1 (1996), pp. 1–40.

[185] Wolfgang Maass and Anthony M Zador. "Dynamic stochastic synapses as computational units". In: *Neural Computation* 11.4 (1999), pp. 903–917.

[186] David MacNeil and Chris Eliasmith. "Fine-tuning and the stability of recurrent neural networks". In: *PloS one* 6.9 (2011), e22885.

[187] Misha Mahowald. "The silicon retina". In: *An Analog VLSI System for Stereoscopic Vision*. Springer, 1994, pp. 4–65.

[188] Zachary F Mainen and Terrence J Sejnowski. "Influence of dendritic structure on firing pattern in model neocortical neurons". In: *Nature* 382.6589 (1996), pp. 363–366.

[189] Abed AlFatah Mansour et al. "An in vivo model of functional and vascularized human brain organoids". In: *Nature Biotechnology* 36.5 (2018), pp. 432–441.

[190] Henry Markram. "The Blue Brain project". In: *Nature Reviews Neuroscience* 7.2 (2006), pp. 153–160.

[191] Henry Markram. "The human brain project". In: *Scientific American* 306.6 (2012), pp. 50–55.

[192] Henry Markram et al. "Reconstruction and simulation of neocortical microcircuitry". In: *Cell* 163.2 (2015), pp. 456–492.

[193] David Marr. *Vision: A computational investigation into the human representation and processing of visual information.* MIT Press, 2010.

[194] Rebecca M Marton and Sergiu P Pașca. "Organoid and assembloid technologies for investigating cellular crosstalk in human brain development and disease". In: *Trends in Cell Biology* 30.2 (2020), pp. 133–143.

[195] Richard H Masland. "The fundamental plan of the retina". In: *Nature Neuroscience* 4.9 (2001), pp. 877–886.

[196] Christian G Mayr and Johannes Partzsch. "Rate and pulse based plasticity governed by local synaptic state variables". In: *Frontiers in Synaptic Neuroscience* 2 (2010), p. 33.

[197] Warren S McCulloch and Walter Pitts. "A logical calculus of the ideas immanent in nervous activity". In: *The Bulletin of Mathematical Biophysics* 5.4 (1943), pp. 115–133.

[198] Lisa M McTeague et al. "Identification of common neural circuit disruptions in cognitive control across psychiatric disorders". In: *American Journal of Psychiatry* 174.7 (2017), pp. 676–685.

[199] Carver Mead. *Analog VLSI and Neural Systems.* Addison-Wesley Longman Publishing Co., Inc., 1989.

[200] Carver Mead. "How we created neuromorphic engineering". In: *Nature Electronics* 3.7 (2020), pp. 434–435.

[201] Carver Mead. "Neuromorphic electronic systems". In: *Proceedings of the IEEE* 78.10 (1990), pp. 1629–1636.

[202] Carver Mead and Lynn Conway. *Introduction to VLSI systems.* Vol. 1080. Addison-Wesley Reading, MA, 1980.

[203] Carver Mead and Mohammed Ismail. *Analog VLSI Implementation of Neural Systems.* Vol. 80. Springer Science & Business Media, 2012.

[204] Adnan Mehonic et al. "Memristors–from In-memory computing, Deep Learning Acceleration, Spiking Neural Networks, to the Future of Neuromorphic and Bio-inspired Computing". In: *arXiv preprint arXiv:2004.14942* (2020).

[205] Paul A Merolla et al. "A million spiking-neuron integrated circuit with a scalable communication network and interface". In: *Science* 345.6197 (2014), pp. 668–673.

[206] Paul Merolla and Kwabena A Boahen. "A recurrent model of orientation maps with simple and complex cells". In: *Advances in Neural Information Processing Systems.* 2004, pp. 995–1002.

[207] Paul Merolla et al. "A digital neurosynaptic core using embedded crossbar memory with 45pJ per spike in 45nm". In: *2011 IEEE Custom Integrated Circuits Conference (CICC).* IEEE. 2011, pp. 1–4.

[208] Claude Meunier and Idan Segev. "Playing the Devil's advocate: is the Hodgkin–Huxley model useful?" In: *Trends in Neurosciences* 25.11 (2002), pp. 558–563.

[209] Petar Milin et al. "Keeping it simple: Implementation and performance of the proto-principle of adaptation and learning in the language sciences". In: *arXiv preprint arXiv:2003.03813* (2020).

[210] Kenneth D Miller, Joseph B Keller, and Michael P Stryker. "Ocular dominance column development: analysis and simulation". In: *Science* 245.4918 (1989), pp. 605–615.

[211] Yuki Miura et al. "Generation of human striatal organoids and cortico-striatal assembloids from human pluripotent stem cells". In: *Nature Biotechnology* 38.12 (2020), pp. 1421–1430.

[212] Cristopher Moore. "Unpredictability and undecidability in dynamical systems". In: *Physical Review Letters* 64.20 (1990), p. 2354.

[213] Benjamin Morcos et al. "Implementing nef neural networks on embedded fpgas". In: *2018 International Conference on Field-Programmable Technology (FPT).* IEEE. 2018, pp. 22–29.

[214] Ryan D Morrie and Marla B Feller. "A dense starburst plexus is critical for generating direction selectivity". In: *Current Biology* 28.8 (2018), pp. 1204–1212.

[215] Alessandro Mortara, Eric A Vittoz, and Philippe Venier. “A communication scheme for analog VLSI perceptive systems”. In: *IEEE Journal of Solid-State Circuits* 30.6 (1995), pp. 660–669.

[216] Eilif Muller et al. “Python in neuroscience”. In: *Frontiers in Neuroinformatics* 9 (2015), p. 11.

[217] Alexander Neckar et al. “Braindrop: A mixed-signal neuromorphic architecture with a dynamical systems-based programming model”. In: *Proceedings of the IEEE* 107.1 (2018), pp. 144–164.

[218] John Nickolls and William J Dally. “The GPU computing era”. In: *IEEE Micro* 30.2 (2010), pp. 56–69.

[219] Chuanxin M Niu et al. “Neuromorphic Model of Reflex for Realtime Human-Like Compliant Control of Prosthetic Hand”. In: *Annals of Biomedical Engineering* (2020), pp. 1–16.

[220] Max Nolte et al. “Impact of higher order network structure on emergent cortical activity”. In: *Network Neuroscience* 4.1 (2020), pp. 292–314.

[221] Erkki Oja. “Simplified neuron model as a principal component analyzer”. In: *Journal of Mathematical Biology* 15.3 (1982), pp. 267–273.

[222] John D Owens et al. “GPU computing”. In: *Proceedings of the IEEE* 96.5 (2008), pp. 879–899.

[223] Ruchi Parekh and Giorgio A Ascoli. “Neuronal morphology goes digital: a research hub for cellular and system neuroscience”. In: *Neuron* 77.6 (2013), pp. 1017–1038.

[224] Sergiu P Paşca. “Assembling human brain organoids”. In: *Science* 363.6423 (2019), pp. 126–127.

[225] Edwin Pednault et al. “Leveraging secondary storage to simulate deep 54-qubit sycamore circuits”. In: *arXiv preprint arXiv:1910.09534* (2019).

[226] Bruno U Pedroni et al. “Memory-efficient synaptic connectivity for spike-timing-dependent plasticity”. In: *Frontiers in Neuroscience* 13 (2019), p. 357.

[227] Giovanni Pezzulo and Michael Levin. “Top-down models in biology: explanation and control of complex living systems above the molecular level”. In: *Journal of The Royal Society Interface* 13.124 (2016), p. 20160555.

[228] Baingio Pinna, Gavin Brelstaff, and Lothar Spillmann. "Surface color from boundaries: a new 'watercolor' illusion". In: *Vision Research* 41.20 (2001), pp. 2669–2676.

[229] Michael L Platt and Paul W Glimcher. "Neural correlates of decision variables in parietal cortex". In: *Nature* 400.6741 (1999), pp. 233–238.

[230] Mu-ming Poo et al. "China brain project: basic neuroscience, brain diseases, and brain-inspired computing". In: *Neuron* 92.3 (2016), pp. 591–596.

[231] Christoph Posch et al. "Retinomorphic event-based vision sensors: bioinspired cameras with spiking output". In: *Proceedings of the IEEE* 102.10 (2014), pp. 1470–1484.

[232] Manu Prakash and Neil Gershenfeld. "Microfluidic bubble logic". In: *Science* 315.5813 (2007), pp. 832–835.

[233] Robert Preissl et al. "Compass: A scalable simulator for an architecture for cognitive computing". In: *SC'12: Proceedings of the International Conference on High Performance Computing, Networking, Storage and Analysis.* IEEE. 2012, pp. 1–11.

[234] Economist Quarterly. *After Moore's law.* 2016.

[235] R Quian Quiroga et al. "Invariant visual representation by single neurons in the human brain". In: *Nature* 435.7045 (2005), pp. 1102–1107.

[236] Wilfrid Rall. "Core conductor theory and cable properties of neurons". In: *Comprehensive Physiology* (2011), pp. 39–97.

[237] Vilayanur S Ramachandran. "Blind spots". In: *Scientific American* 266.5 (1992), pp. 86–91.

[238] Tommaso Ranzani et al. "Increasing the Dimensionality of Soft Microstructures through Injection-Induced Self-Folding". In: *Advanced Materials* 30.38 (2018), p. 1802739.

[239] Daniel Rasmussen. "NengoDL: Combining deep learning and neuromorphic modelling methods". In: *Neuroinformatics* 17.4 (2019), pp. 611–628.

[240] Daniel Rasmussen and Chris Eliasmith. "A spiking neural model applied to the study of human performance and cognitive decline on Raven's Advanced Progressive Matrices". In: *Intelligence* 42 (2014), pp. 53–82.

[241] Michael W Reimann et al. "A null model of the mouse whole-neocortex micro-connectome". In: *Nature Communications* 10.1 (2019), pp. 1–16.

[242] Paul Rendell. "Turing universality of the game of life". In: *Collision-based Computing*. Springer, 2002, pp. 513–539.

[243] Robert A Rescorla. "Pavlovian conditioning: It's not what you think it is." In: *American Psychologist* 43.3 (1988), p. 151.

[244] Oliver Rhodes et al. "Real-time cortical simulation on neuromorphic hardware". In: *Philosophical Transactions of the Royal Society A* 378.2164 (2020), p. 20190160.

[245] Oliver Rhodes et al. "sPyNNaker: a software package for running PyNN simulations on SpiNNaker". In: *Frontiers in Neuroscience* 12 (2018), p. 816.

[246] Frank Rosenblatt. "The perceptron: a probabilistic model for information storage and organization in the brain." In: *Psychological Review* 65.6 (1958), p. 386.

[247] Arnd Roth and Mark CW van Rossum. "Modeling synapses". In: *Computational Modeling Methods for Neuroscientists* 6 (2009), pp. 139–160.

[248] Kaushik Roy, Akhilesh Jaiswal, and Priyadarshini Panda. "Towards spike-based machine intelligence with neuromorphic computing". In: *Nature* 575.7784 (2019), pp. 607–617.

[249] Adam J Ruben and Laura F Landweber. "The past, present and future of molecular computing". In: *Nature Reviews Molecular Cell Biology* 1.1 (2000), pp. 69–72.

[250] Bodo Rueckauer et al. "Conversion of continuous-valued deep networks to efficient event-driven networks for image classification". In: *Frontiers in Neuroscience* 11 (2017), p. 682.

[251] Thomas Rueckes et al. "Carbon nanotube-based nonvolatile random access memory for molecular computing". In: *Science* 289.5476 (2000), pp. 94–97.

[252] Joao Sacramento et al. "Dendritic error backpropagation in deep cortical microcircuits". In: *arXiv preprint arXiv:1801.00062* (2017).

[253] Paul-Antoine Salin and Jean Bullier. "Corticocortical connections in the visual system: structure and function". In: *Physiological Reviews* 75.1 (1995), pp. 107–154.

[254] Georg Seelig et al. "Enzyme-free nucleic acid logic circuits". In: *Science* 314.5805 (2006), pp. 1585–1588.

[255] Biswa Sengupta and Martin B Stemmler. "Power consumption during neuronal computation". In: *Proceedings of the IEEE* 102.5 (2014), pp. 738–750.

[256] Teresa Serrano-Gotarredona et al. "STDP and STDP variations with memristors for spiking neuromorphic learning systems". In: *Frontiers in Neuroscience* 7 (2013), p. 2.

[257] Lei Shi et al. "Transgenic rhesus monkeys carrying the human MCPH1 gene copies show human-like neoteny of brain development". In: *National Science Review* 6.3 (2019), pp. 480–493.

[258] Harel Z Shouval, Samuel S-H Wang, and Gayle M Wittenberg. "Spike timing dependent plasticity: a consequence of more fundamental learning rules". In: *Frontiers in Computational Neuroscience* 4 (2010), p. 19.

[259] Shepard Siegel and Lorraine G Allan. "The widespread influence of the Rescorla-Wagner model". In: *Psychonomic Bulletin & Review* 3.3 (1996), pp. 314–321.

[260] Hava T Siegelmann and Eduardo D Sontag. "On the computational power of neural nets". In: *Journal of Computer and System Sciences* 50.1 (1995), pp. 132–150.

[261] Joshua H Singer and Jeffrey S Diamond. "Vesicle depletion and synaptic depression at a mammalian ribbon synapse". In: *Journal of Neurophysiology* 95.5 (2006), pp. 3191–3198.

[262] Michael Sipser. *Introduction to the Theory of Computation*. Cengage Learning, 2012.

[263] Janis Smits et al. "Two-dimensional nuclear magnetic resonance spectroscopy with a microfluidic diamond quantum sensor". In: *Science Advances* 5.7 (2019), eaaw7895.

[264] Niranjan Srinivas et al. "Enzyme-free nucleic acid dynamical systems". In: *Science* 358.6369 (2017).

[265] Ian Stewart. "The Lorenz attractor exists". In: *Nature* 406.6799 (2000), pp. 948–949.

[266] Terrence C Stewart. "A technical overview of the neural engineering framework". In: *University of Waterloo* (2012).

[267] Terrence C Stewart and Chris Eliasmith. "Large-scale synthesis of functional spiking neural circuits". In: *Proceedings of the IEEE* 102.5 (2014), pp. 881–898.

[268] Terrence Stewart, Feng-Xuan Choo, and Chris Eliasmith. "Spaun: A perception-cognition-action model using spiking neurons". In: *Proceedings of the Annual Meeting of the Cognitive Science Society*. Vol. 34. 34. 2012.

[269] Klaus M Stiefel and Terrence J Sejnowski. "Mapping function onto neuronal morphology". In: *Journal of Neurophysiology* 98.1 (2007), pp. 513–526.

[270] Dmitri B Strukov et al. "The missing memristor found". In: *Nature* 453.7191 (2008), pp. 80–83.

[271] Martin Sundermeyer, Ralf Schlüter, and Hermann Ney. "LSTM neural networks for language modeling". In: *Thirteenth Annual Conference of the International Speech Communication Association*. 2012.

[272] Aboozar Taherkhani et al. "A review of learning in biologically plausible spiking neural networks". In: *Neural Networks* 122 (2020), pp. 253–272.

[273] Amirhossein Tavanaei et al. "Deep learning in spiking neural networks". In: *Neural Networks* 111 (2019), pp. 47–63.

[274] Stefan Theil. "Why the Human Brain Project went wrong—and how to fix it". In: *Scientific American* 313.4 (2015), pp. 36–42.

[275] William Thomson. "III. On the theory of the electric telegraph". In: *Proceedings of the Royal Society of London* 7 (1856), pp. 382–399.

[276] Simon J Thorpe. "Spike arrival times: A highly efficient coding scheme for neural networks". In: *Parallel Processing in Neural Systems* (1990), pp. 91–94.

[277] Todd Thorsen, Sebastian J Maerkl, and Stephen R Quake. "Microfluidic large-scale integration". In: *Science* 298.5593 (2002), pp. 580–584.

[278] Anupama J Thubagere et al. "A cargo-sorting DNA robot". In: *Science* 357.6356 (2017).

[279] Michael W Toepke, Vinay V Abhyankar, and David J Beebe. "Microfluidic logic gates and timers". In: *Lab on a Chip* 7.11 (2007), pp. 1449–1453.

[280] Dardo Tomasi, Gene-Jack Wang, and Nora D Volkow. "Energetic cost of brain functional connectivity". In: *Proceedings of the National Academy of Sciences* 110.33 (2013), pp. 13642–13647.

[281] James M Tour and Tao He. "The fourth element". In: *Nature* 453.7191 (2008), pp. 42–43.

[282] James M Tour et al. "Nanocell logic gates for molecular computing". In: *IEEE Transactions on Nanotechnology* 1.2 (2002), pp. 100–109.

[283] Elishai Ezra Tsur. "Computer-Aided Design of Microfluidic Circuits". In: *Annual Review of Biomedical Engineering* 22 (2020).

[284] Elishai Ezra Tsur. "Data Models in Neuroinformatics". In: *Bioinformatics in the Era of Post Genomics and Big Data* (2018), p. 133.

[285] Elishai Ezra Tsur and Michal Rivlin-Etzion. "Neuromorphic implementation of motion detection using oscillation interference". In: *Neurocomputing* 374 (2020), pp. 54–63.

[286] Alan Mathison Turing. *Intelligent Machinery.* 1948.

[287] Alan Mathison Turing. "On computable numbers, with an application to the Entscheidungsproblem". In: *J. of Math* 58.345-363 (1936), p. 5.

[288] David C Van Essen et al. "The Human Connectome Project: a data acquisition perspective". In: *Neuroimage* 62.4 (2012), pp. 2222–2231.

[289] André Van Schaik. "Building blocks for electronic spiking neural networks". In: *Neural Networks* 14.6-7 (2001), pp. 617–628.

[290] HFJM Van Tuijl and Emanuel Laurens Jan Leeuwenberg. "Neon color spreading and structural information measures". In: *Perception & Psychophysics* 25.4 (1979), pp. 269–284.

[291] Benjamin Villalonga et al. "Establishing the quantum supremacy frontier with a 281 pflop/s simulation". In: *Quantum Science and Technology* 5.3 (2020), p. 034003.

[292] Anna L Vlasits et al. "A role for synaptic input distribution in a dendritic computation of motion direction in the retina". In: *Neuron* 89.6 (2016), pp. 1317–1330.

[293] Aaron R Voelker and Chris Eliasmith. *Controlling the Semantic Pointer Architecture with deterministic automata and adaptive symbolic associations.* Technical report, Centre for Theoretical Neuroscience, Waterloo, ON, 2014.

[294] Aaron Russell Voelker. "A solution to the dynamics of the prescribed error sensitivity learning rule". In: *Centre for Theoretical Neuroscience. Tech. Rep.,* Waterloo, ON. 2015.

[295] Aaron Russell Voelker. "Dynamical systems in spiking neuromorphic hardware". In: *Technical report, Centre for Theoretical Neuroscience, Waterloo,* ON, 2019.

[296] Aaron Voelker, Ivana Kajić, and Chris Eliasmith. "Legendre Memory Units: Continuous-Time Representation in Recurrent Neural Networks". In: *Advances in Neural Information Processing Systems.* 2019, pp. 15570–15579.

[297] Vitaly Volpert. *Elliptic Partial Differential Equations: Volume 2: Reaction-Diffusion Equations.* Vol. 104. Springer, 2014.

[298] Fei Wang et al. "Implementing digital computing with DNA-based switching circuits". In: *Nature Communications* 11.1 (2020), pp. 1–8.

[299] Yequan Wang et al. "Attention-based LSTM for aspect-level sentiment classification". In: *Proceedings of the 2016 Conference on Empirical Methods in Natural Language Processing.* 2016, pp. 606–615.

[300] Heinz Wässle. "Parallel processing in the mammalian retina". In: *Nature Reviews Neuroscience* 5.10 (2004), pp. 747–757.

[301] Paul John Werbos. *The Roots of Backpropagation: From Ordered Derivatives to Neural Networks and Political Forecasting.* Vol. 1. John Wiley & Sons, 1994.

[302] James CR Whittington and Rafal Bogacz. "An approximation of the error backpropagation algorithm in a predictive coding network with local hebbian synaptic plasticity". In: *Neural Computation* 29.5 (2017), pp. 1229–1262.

[303] Bernard Widrow and Marcian E Hoff. *Adaptive switching circuits.* 960 IRE WESCON Convention Record, 1960, pp. 96–104.

[304] Bernard Widrow et al. *Adaptive adaline Neuron Using Chemical memistors.* Stanford Electronics Laboratories Technical Report, 1553-2, 1960.

[305] Leanne M Williams. "Precision psychiatry: a neural circuit taxonomy for depression and anxiety". In: *The Lancet Psychiatry* 3.5 (2016), pp. 472–480.

[306] Theodore M Wong et al. *Ten to power 14.* IBM Research Report RJ10502, Volume ALM1211-004, 2013.

[307] Damien Woods et al. "Diverse and robust molecular algorithms using reprogrammable DNA self-assembly". In: *Nature* 567.7748 (2019), pp. 366–372.

[308] Guangyu Robert Yang and Xiao-Jing Wang. "Artificial neural networks for neuroscientists: A primer". In: *arXiv preprint arXiv:2006.01001* (2020).

[309] J Joshua Yang et al. "Memristive switching mechanism for metal/oxide/metal nanodevices". In: *Nature Nanotechnology* 3.7 (2008), pp. 429–433.

[310] Yong Yu et al. "A review of recurrent neural networks: LSTM cells and network architectures". In: *Neural Computation* 31.7 (2019), pp. 1235–1270.

[311] Rafael Yuste. "From the neuron doctrine to neural networks". In: *Nature Reviews Neuroscience* 16.8 (2015), pp. 487–497.

[312] Kareem A Zaghloul and Kwabena Boahen. "A silicon retina that reproduces signals in the optic nerve". In: *Journal of Neural Engineering* 3.4 (2006), p. 257.

[313] Yuval Zaidel et al. "Neuromorphic NEF-based inverse kinematics and PID control". In: *Frontiers in Neurorobotics* 15 (2021).

[314] Minglu Zhu, Tianyiyi He, and Chengkuo Lee. "Technologies toward next generation human machine interfaces: From machine learning enhanced tactile sensing to neuromorphic sensory systems". In: *Applied Physics Reviews* 7.3 (2020), p. 031305.

[315] Shay Zweig et al. "Representation of color surfaces in V1: edge enhancement and unfilled holes". In: *Journal of Neuroscience* 35.35 (2015), pp. 12103–12115.

Index